OXFORD MEDICAL PUBLICATIONS

Problem Drinking

Problem Drinking

Second Edition

NICK HEATHER
National Drug and Alcohol Research Centre
University of New South Wales, Australia

and

IAN ROBERTSON
Astley Ainslie Hospital, Edinburgh

Oxford   New York   Tokyo
OXFORD UNIVERSITY PRESS
1989

Oxford University Press, Walton Street, Oxford OX2 6DP

Oxford New York Toronto
Delhi Bombay Calcutta Madras Karachi
Petaling Jaya Singapore Hong Kong Tokyo
Nairobi Dar es Salaam Cape Town
Melbourne Auckland

and associated companies in
Berlin Ibadan

Oxford is a trade mark of Oxford University Press

Published in the United States
by Oxford University Press, New York

© Nick Heather and Ian Robertson, 1989

First published by Penguin Books, Ltd, 1985
Second edition, 1989

British Library Cataloguing in Publication Data
Heather, Nick, 1938–
Problem drinking.—2nd ed (Oxford medical publications)
1. Alcoholism. Psychosocial aspects
I. Title II. Robertson, Ian, 1951–
362.2'92
ISBN 0-19-261874-1

Library of Congress Cataloging in Publication Data
Heather, Nick.
Problem drinking. (Oxford medical publications)
Bibliography: Includes index.
1. Alcoholism—Etiology. 2. Alcoholism—Psychological
aspects 3. Alcoholism—Social aspects. I. Robertson.
Ian, 1951– . II. Title. III. Series. [DNLM:
1. Alcohol Drinking. 2. Alcoholism—psychology.
3. Social problems. WM 274 H441p]
RC565.H32 1989 362.29'2 89-8582
ISBN 0-19-261874-1 (pbk.)

Set by CentraCet, Cambridge
Printed in Great Britain by
The Guernsey Press Co. Ltd.,
Guernsey, Channel Islands.

Contents

Preface to the second edition

In this second edition of *Problem drinking*, apart from correcting mistakes and updating statistics and references, we have tried to cover the major developments in the field since the book was written four or five years ago.

In Chapter 2, the brief historical background to the subject is extended to cover a development which has increased considerably in importance since the first edition. This is the expansion of the 'alcoholism treatment industry', particularly in the United States, in which, contrary to the indications of all the available evidence on treatment effectiveness, expensive in-patient regimes based largely on the disease ideology and methods of Alcoholics Anonymous are being aggressively marketed and consumed. Whether this will save the disease theory in the long run is another matter.

Another section requiring revision and expansion was the review of the 'controlled-drinking controversy' in Chapter 4. Although there have been several important studies published in this area since the first edition, by no means all unfavourable to the case for controlled drinking, one thing has not changed— the blinkered and irrational attitudes to this issue that are apparent in many quarters.

The remaining major changes occur in the discussion of the practical implications of the social learning paradigm of problem drinking in Chapter 9. In particular, since we first wrote, there has been much progress in exploring the possibilities of community-based brief interventions aimed at early recognition and modification of problem drinking and the chapter has been enlarged to take account of this. Chapter 9 and parts of the

Epilogue have also been changed to allow descriptions of recent social and political developments in Britain.

As the last sentence implies, despite the fact that one of the authors has subsequently moved to work in Australia, we have not changed the book's focus on treatment and prevention in the British context. To have done so would have meant a different book. However, as with the first edition, we are confident that the ideas and evidence reviewed have an international relevance.

We are aware of several criticisms that were made of the first edition and have paid attention to them. One was that, although aimed partly at the general reader, the book was too difficult for this readership and was much more attuned to those studying or professionally involved in the subject. This surprised us because we had deliberately tried to avoid unnecessary technical terms and arcane topics. The level of writing of this second edition has not been altered and we continue to hope that the general reader seeking a serious introduction to the issues involved, rather than an over-simplified or trivialized version of them, will find the book a rewarding experience.

One other criticism deserves special mention here. This is the notion that the promotion of the new approach to problem drinking has nothing to do with 'paradigm change' but is merely part of a conspiracy among psychologists to acquire 'ownership' of the field from their medical colleagues. While it is true that paradigm changes typically involve social and political conflict between scientists, this is a battle of ideas, not of professional interests. The absurdity of thinking otherwise can be simply demonstrated from the fact that some of the most distinguished contributors to the learning theory of drug dependence are themselves medically qualified.

No one could deny that, among those working in this area, there is sometimes a struggle for dominance between psychology and medicine, but to confuse this with the scientific debate represents a particularly unhelpful form of reductionism.

Besides, to try to dismiss the case for the learning approach in this way is an excellent example of the classic logical fallacy of the argument *ad hominem*—the attempt to refute someone's position by impugning their motives in supporting it. In our view, the idea that the advancement of the social learning paradigm is nothing more than a self-interested bid by psychologists to 'take over' the problem drinking field is neither a useful nor a serious contribution to the debate.

Sydney   N.H.
Edinburgh   I.R.
February 1989

1 Introduction

This book is written for both the general reader and the specialist. It aims to describe and summarize the important changes which have taken place over the last fifteen to twenty years in the scientific understanding of problem drinking. But although written partly for the general reader, the book is not simply another general introduction to the subject of alcoholism.[1] It has a more specific purpose than that. The main intention is to present a sustained argument, one which derives from the scientific advances of the last few years but which, as far as we are aware, has not been made available to the general public before.

The argument of the book is that *problem drinking is not best regarded as a disease but as a learned behavioural disorder*. This argument is far from new. It is probably fair to say that the majority of experts in the field in Britain would now reject the idea that there is something which can sensibly be called a disease of alcoholism. If this seems surprising to the reader, it is only because there exists such a large gap between contemporary scientific understanding and what the public is usually exposed to in popular accounts of the subject. It is this gap the book aims to fill. When the lay person does catch a glimpse of the new understanding, it often leads only to confusion, since the insight is rarely substantiated by a more rounded picture of recent findings and theoretical developments. It is this confusion the book aims to dispel.

In arguing that problem drinking is not a disease, it is possible that we ourselves will be accused of spreading confusion. A paradoxical effect of the gap in communication between specialist and non-specialist is that, at precisely the time when the disease view of problem drinking is being abandoned and regarded as hopelessly outmoded by many professional workers, it is still thought of as a mark of liberal and enlightened opinion among the general public. Thus, it is still common to hear of attempts to persuade people to accept that alcoholism really *is* a disease. Apart from any scientific merits it may or may not possess, the major effect intended by supporters of the disease perspective has been to persuade the public at large that the alcoholic is not responsible for his troublesome behaviour and therefore cannot be blamed for it. Alcoholics 'can't help' behaving the way they do because they are suffering from a disease; hence they should not be punished but should receive help and treatment. This humanitarian response to alcoholism is now very familiar and over the last forty years (as we shall see in Chapter 2) a serious and concerted attempt has been made to persuade the public to adopt it.

However, the disease view of alcoholism is seen as enlightened simply because the only apparent alternative is to go back to blaming and punishing alcoholics. Those who would accuse us of spreading confusion would no doubt suggest that such a reversion to punitive attitudes is the inevitable consequence of telling people that alcoholism is not a disease. With this possibility in mind, therefore, it is essential to state clearly for the first time what will be repeated in various ways throughout the book and what perhaps represents its principal, single message: *to deny that problem drinking is a disease is not to deny that problem drinkers should receive compassion and help*. Thus a major task for the book is to explain how it is possible to reject the idea that alcoholics are 'sick' without at the same time implying that they are 'bad'.

A scientific revolution for alcoholism?

If it is not a return to a moralistic and punitive attitude, what then is the alternative to the disease view of alcoholism which is being recommended? The broad answer is the idea that problem drinking is a consequence of the way the individual has learned to drink alcohol, this learning being influenced in turn by the social and cultural context in which drinking has occurred. This change from a disease to a learning perspective has occupied the alcoholism research and treatment community for the last decade and has been hailed by some as a scientific revolution—or, in more technical jargon, 'a paradigm change'. Let us explore what this means.

Originally published in 1962, a book by the historian of science, Thomas S. Kuhn, entitled *The structure of scientific revolutions*,[2] has proved very influential among scientists of all kinds. Kuhn's chief concern is with the way in which scientific knowledge grows. He rejects the received 'common-sense' idea that it does so in a continuous and straightforward way—as, for example, in the naïve and quite erroneous view of science which sees it as a gradual piling-up of more and more 'facts' until a perfect understanding is eventually reached. In contrast to this, Kuhn shows that science proceeds by a succession of discontinuous jumps which radically alter the basic assumptions of the discipline in question. In between these jumps occur periods of 'normal science', in which the important problems to be solved, the methods by which solutions to these problems are sought, and, indeed, the very standards by which correct solutions are recognized are all agreed on by the relevant scientific community. This conceptual and practical framework, which serves to regulate the science during its normal periods, is called by Kuhn a 'scientific paradigm'; the principal activity of scientists during these normal periods takes the form of puzzle-solving exercises, rather than the more glamorous pursuit of uncharted mysteries with which science is often incorrectly associated.

Sooner or later, however, these cosy periods of paradigm-regulated activity are disrupted by a completely novel finding, one which may happen quite by chance, which cannot be predicted from the existing paradigm or be made to find a place within it. These 'anomalies', as Kuhn calls them, are the only true discoveries of which science is capable. Preoccupation with them leads to a state of crisis in which the technical, puzzle-solving activity of normal science breaks down and is replaced by a re-examination, often involving considerable bitterness and controversy, of the fundamental assumptions which have up till then been in force. The response of the scientific community is typically one of polarization, with some defending the old paradigm and others (usually the younger members) urging its replacement by the new one. Eventually, the outlines of the new paradigm gradually emerge and the discipline enters its next normal period, with its own key theoretical problems, unique research methods, and agreed rules for valid findings.

We have devoted some space to the notion of paradigm change because it has been claimed that this is now happening in the field of alcohol studies with respect to the change from a disease to a learning model. On the other hand, some commentators have argued that this transition is an example of the development from the stage of pre-scientific activity to the first formation of a science proper. According to this view, the disease perspective of alcoholism constitutes a 'folk science', relying on hearsay and anecdotal material and on inductive styles of reasoning in which sweeping claims are made on the basis of limited evidence. This is opposed to the methods of science proper, in which the collection of data is restricted to controlled conditions and deductive reasoning is employed in order to arrive at testable hypotheses.

Although it clearly has its limitations, the application of the concept of paradigm change to the problem-drinking field does have certain attractions. One is that there is such an obvious candidate for the position of the Kuhnian anomaly—the single, unprecedented finding which subverts the old paradigm and leads to its ultimate downfall. In this case the finding was the observation of D. L. Davies, reported in 1962, that seven 'alcohol addicts' he had followed up at least seven years after being discharged from treatment at the Maudsley Hospital, London, had been able to return to a normal pattern of harm-free drinking (see Chapter 4). There are many other findings which have become embarrassing to the disease theory, but it is Davies's discovery of 'resumed normal drinking', which has now been replicated many times over, that has proved hardest for it to absorb.

Mention of the finding of resumed normal drinking provides an opportunity to introduce a warning which will be repeated later in the book. In attacking the disease theory of alcoholism and in stressing the fact that some problem drinkers are able to drink again in a normal manner, we are definitely *not* recommending that all alcoholics should attempt controlled drinking. It is necessary to be absolutely clear about this. As we shall see, the evidence suggests that the majority of seriously dependent problem drinkers are still more likely to succeed in solving their problem by opting for total abstinence from alcohol. *Nothing we write in this book should be interpreted as urging alcoholics to reject abstinence as a goal of treatment. Equally, nothing we say here should be construed as encouraging those former problem drinkers who have achieved a contented life through total abstinence to return to drinking.* Any successful solution to a drinking problem is far too valuable to be tampered with.

Besides the hostility generated by most scientific disputes, there is an extra ingredient which may be unique to the alcoholism field. This stems from the fact that many people closely involved in treatment and education in the area are themselves former problem drinkers or, in terms which as affiliates of Alcoholics Anonymous they would no doubt prefer, 'recovering alcoholics', who are now committed to total and lifelong abstention. The involvement of such people extends from the grass roots of providing counselling services up to the management of national statutory and voluntary agencies, with the power to grant or withhold financial support and to influence the formation of national policy. This tendency is more pronounced in the United States but is still true of Britain. We will comment further on this influence when we come to describe the history and contemporary standing of Alcoholics Anonymous in Chapter 2. Suffice it to point out here that many individuals involved in the alcoholism field have a vested emotional interest in the disease theory of alcoholism.

Why problem drinking?

So far in the book, the terms 'problem drinking' and 'alcoholism' have been used more or less interchangeably. It is time these terms were clarified.

Part of the difficulty with this is that 'problem drinking' has been used in at least two very different senses. One common use of the term is in connection with harmful drinking behaviour which is somehow regarded as less serious than other harmful drinking which deserves the title of alcoholism. The distinction is usually made by referring to the occurrence in alcoholism of physical withdrawal symptoms, such as delirium tremens (the DTs) and 'the shakes', if no alcohol is consumed for some time. Thus alcoholism here is more or less equivalent to physical addiction to alcohol, in the same way as drug addicts are thought to be addicted to heroin. In this use, therefore, problem drinking does not involve physical addiction and is carried on for psychological and emotional reasons and not because of a need to get rid of unpleasant withdrawal symptoms, as in alcoholism proper. This kind of distinction was first made by E. M. Jellinek, undoubtedly the leading authority the field of alcohol studies has yet known, in a book called *The disease concept of alcoholism* published in 1960.[3] In a classic description of different species of alcoholism, Jellinek distinguished between gamma and delta alcoholisms, which involved physical addiction and which were rightfully called diseases, and alpha alcoholism, which was based only on the need to relieve emotional pain and was not a true disease. An alternative name for alpha alcoholism was 'problem drinking'.

However, *this is not the meaning of problem drinking which will be used in this book*. Rather, we will use the term in the most general sense possible. If pushed for a formal definition, we would appeal to one proposed by the sociologist Don Cahalan in 1970: 'Problem drinking is repetitive use of beverage alcohol causing physical, psychological or social harm to the drinker or to others'.[4] Strangely enough, this definition is very similar to a wide-ranging use of the term alcoholism made by Jellinek in *The disease concept of alcoholism*, where it was defined as 'any use of alcoholic beverages that causes any damage to the individual or to society or both'. For convenience, we will often use 'alcoholism' as a synonym for problem drinking in this wide-ranging and all-embracing sense.

But if alcoholism can be used in this very general sense, why bother to use an alternative term like 'problem drinking'? The answer is that when one speaks of alcoholism one is usually interpreted as referring to the *disease* of alcoholism. Since we wish to abandon the disease perspective, it is essential also to abandon the term which is associated with it. In other words, in adopting a fresh term, the reader is invited to make a fresh start in thinking about the problems caused by alcohol.

The alcoholism epidemic

If terms like alcoholism and problem drinking have caused disagreement, there is one thing about which all observers are agreed: there is a lot of it about!

It is hardly possible to lift a newspaper or switch on a television news programme these days without coming across some reference to the rising tide of alcohol problems. To take the most obvious index, the number of hospital admissions for 'alcoholism' or 'alcoholic psychosis' per year in Britain rose from 512 in 1952 to 13 916 in 1982.

Obviously, this massive increase was partly a reflection of changes in diagnostic practices by the medical profession and the great increase in the number of places available in hospital for alcoholic patients. But there has been a real increase of considerable proportions too. This is shown by the fact that during the last decade, when in-patient facilities for alcoholism in the National Health Service have remained roughly constant, the number of admissions with a diagnosis of alcoholism has doubled. Moreover, the rise in alcoholism diagnoses since 1952 has been accompanied by increases in other indicators of alcohol problems over a similar period. For example, convictions for drunkenness went up from twelve per 10 000 of population in 1952 to twenty-six per 10 000 in 1983. Deaths from cirrhosis of the liver have likewise nearly doubled, from below 1500 in 1952 to 2700 in 1984. More recently, convictions for drinking and driving have increased from 23 971 in 1968, when the breath test was introduced, to 113 213 in 1984. Again, this may partly be due to changing police policies but it undoubtedly measures a genuine increase in drunken driving as well. Finally, underlying all these measures of increasing damage and notwithstanding a levelling-off in the late 1970s, there has been a doubling of alcohol consumption from 4.9 litres per adult in 1950 to 9.2 litres in 1984. The largest proportional increases in consumption and in problems since the war have been in women and young people.

Looking at all the different indices of harm, and adding figures for Scotland and Northern Ireland to those for England and Wales, it is possible to make a rough estimate that no less than one million people in the United Kingdom today are experiencing serious problems related in some way to their use of alcohol. This is about four times the number in the early 1960s. On the basis of the figures alone, and imagining for the moment that alcoholism really were an infectious or contagious disease, the word 'epidemic' is not out of place.

Nevertheless, the extent of the problem in Britain should be kept in proportion. As we shall see in Chapter 2, there have been times past when we drank even more, and times when 'the drink question' was a hotter political issue than it is now. Furthermore, although international comparisons are strongly influenced by the kind of measure being employed, Britain appears relatively well off compared with other European countries. For instance, liver cirrhosis mortality rates in the United Kingdom come near the bottom of the European league table and are only a fraction of those recorded in countries like France, Italy, and Austria. At the same time, we consume less than half the alcohol drunk by the French and the rate of increase in our consumption over the last twenty years or so has been considerably less than in some other European nations, most notably West Germany.

But although it is true that things could be worse, it is worth emphasizing that Britain is sharing with the rest of the industrialized world, and with large parts of the non-industrialized world, a real and significant increase in the level of problems related to alcohol. If there is a novel aspect to the post-war increase in alcohol-related problems, it is precisely that this increase is a world-wide phenomenon—a 'pandemic' rather than merely an 'epidemic'.

There is also no doubt that the British government is keenly aware of this increase and concerned about it, but is apparently unwilling to take effective action to tackle the problem. This is shown by the notorous fate of the Think Tank report, which was ordered by the Labour government in 1978 but suppressed by the Conservatives when they came to power in 1979. The background and remarkable history of this report will be described in more detail in Chapter 9 (pp. 309–11).

It will now be evident that issues relating to the best way to describe and understand alcohol problems—for example, whether they are best understood in the language of disease or in some other way—are far from merely academic issues. In particular, it is a question of immediate practical concern how best to explain the dramatic increase in alcohol-related problems since the war. It is our belief that, if society is to work out a rational, humane, and effective policy towards the individual and social harm caused by the misuse of alcohol, it is essential to begin with a proper understanding of the phenomenon and to communicate this understanding as widely as possible. It is in this belief that the book has been written.

Plan of the book

The book is divided into two parts. The first is unapologetically destructive and aims to review all the evidence currently available which has rendered the disease position untenable. It would be foolish to insist that every piece of evidence ever collected goes against the disease view; some research, particularly studies of family inheritance of alcoholism, is apparently in its favour and must be considered carefully by anyone urging the abandonment of the disease perspective. But what we would assert is that, when weighed and compared, the evidence is overwhelmingly unfavourable to a disease theory of alcoholism.

Before considering the recent evidence in detail, however, it is necessary to place the disease perspective in its historical and social context and this we do in Chapter 2. Here, we show that many important aspects of the twentieth-century disease explanation of problem drinking are in fact 'hangovers' from the response to alcohol problems of the previous century. Chapter 2 also contains a brief historical account of the development of that remarkable fellowship of men and women, called Alcoholics Anonymous, together with an assessment of AA's current role in the treatment of drinking problems. The chapter ends with a brief account of some disturbing recent developments.

Having prepared the historical foundations, we clarify in Chapter 3 the several senses in which alcoholism has been described as a disease, with their quite different implications for the alleged causes of the condition. Chapter 4 presents a summary of the research findings that have subverted the disease position and, recognizing that a discussion of its merits should not be confined to research data, Chapter 5 debates the overall advantages and disadvantages of the disease perspective for a rational policy towards alcohol problems.

Having devoted the first part of the book to a critical analysis of the disease perspective, we turn in the second part to its replacement—the social-learning paradigm of problem drinking. Chapter 6 demonstrates that dependence on alcohol is not primarily a consequence of the effects of the drug ethyl alcohol, but is related rather to a range of 'addictive' behaviours which occur without any drug involvement. The chapter continues with an attempt to define problem drinking and a very brief account of the wider theoretical background to the new paradigm in the main principles of social-learning theory.

The next two chapters go on to show how this loose-knit body of psychological principles can be used to explain drinking behaviour, both harmful and harm-free. In Chapter 7, we outline the roles of classical and operant conditioning, and Chapter 8 is an account of the part played by higher-order learning processes. Finally, in Chapter 9, we consider the implications of the social-learning approach for treatment and prevention policy, and argue for a radical rethinking of the response to problem drinking in Britain.

Part 1

Problem drinking
is not a disease

2   The historical context

Writing in 1773, Dr Johnson lamented the passing of the good old days 'when all the decent people in Lichfield got drunk and every night and was not thought the worse of'.[1] Presumably such nostalgia already reflected the emergence of a more well-disciplined and sober age but, even so, on a night in 1782, James Boswell, Johnson's faithful companion and biographer, was found dead drunk in a London gutter.

The eighteenth century in Britain was an era of prodigious drinking of alcohol. At the height of the so-called 'gin epidemic' in 1750, an English population of just over six million managed to finish off eleven million gallons of gin.[2] Just before the turn of the century in 1689, the annual per capita intake of beer alone has been calculated at 832 pints, or an average of 2.3 pints per person per day. When it is remembered that this calculation includes all children and non-drinkers, and that beer in those days was considerably stronger than its modern counterpart, it will be appreciated just how heavy the consumption of the regular drinkers must have been.

Figure 2.1 is a graph of alcohol consumption in Britain from about 1680 to 1975.[3] Consumption is expressed as grammes of pure alcohol per person per day and is shown separately for beer, wine, and spirits. The quantities are taken from HM Customs and Excise records over the years in question and are only rough approximations of amounts of alcohol actually consumed.

In the eighteenth century, over-indulgence was by no means confined to the 'lower orders' but extended, with only a change of preferred beverage, through all levels of society. The household accounts of Sir Robert Walpole in 1733 show that he had

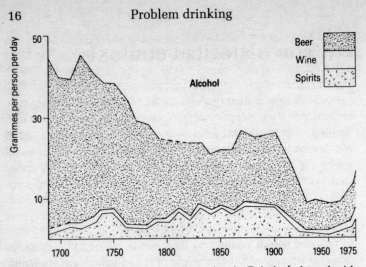

Fig. 2.1. Changes in alcohol consumption in Britain [adapted with permission from Spring, J. A. and Buss, D. H. (1977). Three centuries of alcohol in the British diet. *Nature*, **270**, 567–72].

paid over £1000 to a wine merchant for vintage clarets and burgundies. After entertaining friends at his country estate, and even though only the best French wines were bottled rather than drawn from barrels, Sir Robert returned to the same merchant no less than 540 dozen empty bottles a few months afterwards. The alcoholic excesses of a later Prime Minister, John Carteret, were so conspicuous that his term of office in the 1740s became known as 'the drunken administration'. And at his wedding in 1795, the Prince Regent and future George IV could hardly stand!

Before the start of the eighteenth century, this love of intoxication had been exported to the British colonies. For example, in 1678 in Boston at the funeral of one Mary Norton, the wife of a well-known minister, the mourners drank over fifty gallons of wine. A few years later, at the ordination of the Reverend Edwin Jackson of Woburn, Massachusetts, the guests put away six and a half barrels of cider, twenty-five gallons of wine, two gallons of brandy, and four gallons of rum.[4] We are not told how many guests there were but, whatever their number, such quantities are impressive.

It was in this context of gargantuan appetites for alcohol that the first disease concept of alcoholism arose. It is in this context also that the exaggerated reaction to alcohol of the nineteenth century must be understood.

The Temperance Movement

It is sometimes forgotten today how powerful the Temperance Movement was. Purely in terms of numbers, as early as 1835, one and a half million inhabitants of the United States, out of a total population of thirteen million, had 'signed the pledge' vowing never to touch alcohol again. In the United Kingdom in 1889, the Band of Hope, the temperance organization for juveniles, could claim 16 000 societies and two million members, a figure which had risen to over three million by the end of the century or more than one in four of the relevant age group.

But these figures fail to convey the full reality. In the ninteenth century, the issue of temperance penetrated almost every important aspect of life—respectability and deviance, work and play, personal aspirations and political convictions. It affected the lives of our great-grandparents in a way which is now difficult for us to grasp. Any movement that could succeed in changing the constitution of the United States of America must surely have had tremendous strength.

Attitudes to alcohol before the nineteenth century

It was in the United States that the Temperance Movement began, spreading from there to Britain and other parts of the world. Before we explore its origins in detail, however, it will be useful to set the scene by describing the attitudes to alcohol which preceded it.

In seventeenth- and eighteenth-century Britain and in colonial America, alcohol was accorded a very high esteem. This was an era when beer was regarded as safer than water—not without reason, since the latter was frequently polluted, being drawn in many cases from the same river which served as the local sewer. Moreover, a regular supply of alcohol was regarded as essential for good health, much as the French still regard wine. Indeed, there is evidence that calories from alcohol formed an essential part of the population's energy requirements, while beer made a major contribution to the nutritional content of its diet.[3] In accordance with the high status of drink, the tavern served a vital function as a meeting place in each community and the innkeeper was often numbered among the most prominent of local citizens.

This elevated prestige of alcohol extended right across society and embraced denominational groups like the Puritans. In his celebrated sermon *Wo to drunkards* published in 1673, the great Puritan divine Increase Mather stated unequivocally that 'drink in itself is a good creature of God and to be received with thankfulness'; it was only the abuse of drink and drunkenness which sprang from the Devil. Even so, Puritans often showed compassion for drunkards and attempted to reform rather than castigate them. The image of the Puritan as the unrelenting and unforgiving scourge of any use of alcohol is largely nineteenth-century temperance propaganda.[5]

This does not mean that drunkenness went unpunished before 1800. The range of punishments available to the magistrate included fines, whippings, stocks, and occasional imprisonment, and penalties for repeat offences were sometimes severe. Civil authorities tended to back up the sentences already conferred by the ecclesiastical courts and the wrongdoer was often doubly chastised. Attempts were also made to make the punishment fit the crime. In Elizabethan England, offenders were forced to parade through the streets wearing a 'drunkard's cloak'—a bottomless beer-barrel with holes cut out for the arms—while in the following century in Massachusetts, a culprit was disenfranchised and made to wear a scarlet 'D', for drunkard.

However, drunkenness was not considered a major problem in the eighteenth century. It was implicitly assumed that it could be contained by the normal processes of social control within local communities. There were exceptions to this rule among special classes of the population, and gin drinking by the destitute of London and other towns posed considerable legislative difficulties for successive governments during the first half of the century. But this was seen as a problem confined to the urban poor; it did not impinge on other sectors of society and certainly did not affect the generous appetites of the law-makers themselves. Similarly, drunkenness among Indians and black slaves was the cause of some consternation in colonial America, but these problems were clearly distanced from normal society and the lives of the colonists themselves. Hypocritical they may have been, but they were unable to perceive any serious threat to the social order from alcohol.

Precisely why this changed and why the next century discovered a major social problem where none existed before is mysterious. It cannot have been due to any increase in the amount consumed because no such increase occurred. At the very beginning of the swing towards temperance, there was a reaction specifically directed against spirits, which were of relatively recent origin. Although spirit drinking had been curtailed in Britain, it had continued to increase in America, where the Temperance Movement began. The American equivalent of British gin was at first the 'demon rum' but, towards the end of the eighteenth century, rum was replaced in popularity by home-produced whiskey and, in the opening decades of the nineteenth century, the geographical and population explosion of the new republic was accompanied by a corresponding explosion in whiskey drinking. This undoubtedly contributed to temperance indignation but cannot account for the continuing strength of the movement.

Another possibility is to refer to the much greater complexity of work and interdependence between workers demanded by the new technology of the Industrial Revolution. Frequent imbibing through the day could ease the back-breaking toil of the agricultural labourer without much disruption to the work process, and eighteenth-century employers provided their hands with a generous supply. But the same did not apply to the machine-minder, whose intoxication might well impair levels of production and endanger the safety of others besides himself. In this way, so this explanation goes, drinking became a threat to the economic substructure of the new order. In addition, the great increase in economic and social mobility and the emergence of bewildering changes in moral values which accompanied the break-up of the old order seemed to fragment networks of deference and respect and threaten to swamp existing institutions of social control. In the United States, these changes were made more profound by the lawlessness of the Western frontier and the waves of immigration which periodically swept across the country. Thus drunkenness caused anxiety to conservative interests because it appeared to be a precursor of social chaos.

A different sort of explanation has been proposed by the American sociologist Joseph R. Gusfield.[6] In this theory, it is not economic interests which are of paramount importance but interests in prestige and status within society. Gusfield argues that conspicuous avoidance of alcohol was used by the emergent middle classes of the Industrial Revolution—the skilled mechanics and tradesmen—as a sort of status symbol and a validation of superior social position. Conversely, since heavy drinking became recognized as a mark of disreputable working-class life, the antithetical behaviour of abstinence was used to communicate a rejection of working-class norms and values. Thus, for the working-class person with middle-class aspirations, abstinence became, in Gusfield's phrase, 'a ticket of admission into respectability'. The reason why abstinence was chosen as a symbol was because of its association with the moral qualities of self-control, industriousness and thrift, qualities prized as accompaniments of prosperity. In this way, a display of abstinence demonstrated one's religious adherence and at the same time showed that one possessed the necessary character for success in life.

Beginnings of the Temperance Movement

The precise point at which the Temperance Movement began has been dated at various times but most historians see 1826 as the significant year. This witnessed the publication of Lyman Beecher's *Six sermons on intemperance*, which was to become a classic of temperance literature, and the formation of the American Temperance Society, the first national organization in the field and the first amalgamation of temperance enthusiasts from the various religious denominations. It was probably in 1826, too, that the first pledges against taking intoxicating liquor were made in Boston.

Soon after this, seamen on shore-leave from American vessels began distributing temperance tracts in the port of Liverpool. As in the United States, there had been sporadic and isolated attacks on intemperance before this but no organized body had emerged. The true founder of the movement in Britain was Joseph Livesey, a Preston cheesemonger who, after about 1830, devoted all his considerable talents and energies to the crusade.

It should be pointed out immediately that the character of this crusade was very different from its modern counterpart.

Contrary to the eccentric and conservative image of contemporary teetotalism, temperance supporters in 1830 saw themselves at the forefront of radical protest and reform, believing that science and progress were unmistakably on their side. Many were also enthusiasts for other reforming causes—the abolition of slavery, extension of the suffrage, public education, women's rights, and so forth. But temperance was the leading and most urgent reform; there was little point in liberating, educating, or saving people from a life of crime if they went straight back to the bottle.

It may have been this radical appeal at a time of great social turmoil which partly accounts for the extraordinary early success of the Temperance Movement. Whatever the reason, in both Britain and America it burgeoned in a way which no other broadly popular movement has done before or since.

The emergence of teetotalism and of legislative coercion

Combined with this rapid growth, however, there quickly developed the factions and splits to be expected in any movement. One of the first disagreements concerned the widening of disapproval from just spirits to all alcoholic beverages. Influenced by the events of the eighteenth century, the early reformers had confined their warnings to the unnatural 'spiritous liquors', while continuing to insist that beer was a wholesome and necessary part of the diet. Later and more extreme converts would have none of this, maintaining that alcohol in any form was the first step on the road to perdition.

Related to this dispute was the battle between those who favoured temperance in the literal sense of the word—moderation and avoidance of excess—and those who ridiculed the idea that moderation was possible and argued that 'the first drink' led invariably to insanity or death. On both counts, the more extreme position won easily and, henceforth, the Temperance Movement was aligned almost entirely behind total and lifelong abstention from all forms of alcohol.

There are two versions of how the term 'teetotaller' came about, one for each side of the Atlantic. In America it is claimed that at a meeting in 1827, those who had taken the 'total pledge' were marked with a 'T' in the membership roll. The British version concerns an early zealot named Dicky Turner who tended to stutter when excited. One famous night Turner exclaimed to a packed meeting that he would have none of 'this moderation, botheration pledge', but would be outright 'tee-tee-total for ever and ever'.

The next major split in the ranks concerned the tactics to be employed to achieve the agreed aim of universal abstention. Early temperance workers had believed in the power of 'moral suasion'—in other words, persuading as many people as possible to sign the pledge. In this way, they reasoned, the drink industry would gradually wither away from lack of custom and a dry Utopia would spontaneously arise. However, after the movement's first flush of success had subsided, it appeared to many that these tactics were taking too long to work and more impatient counsels began to prevail. In contrast to moral suasion, 'legislative coercion' was put forward as the key to success and, after the formation of the United Kingdom Alliance for the Suppression of the Traffic in all Intoxicating Liquors in 1853, the legal prohibition of the alcohol industry became the dominant aim of the Temperance Movement.

During these battles there was one source the movement could rely on for recruitment and support—organized religion. It is an exaggeration to say that temperance and Catholicism were incompatible, as the remarkable if temporary success of Father Matthew's crusade in Ireland in the 1840s plainly demonstrates, but it was the Protestant denominations where temperance found its staunchest allies. Moreover, it was the dissenting Protestant churches—Congregationalist, Quaker, Presbyterian, Baptist, and Methodist—which yielded the most energetic supporters. These denominations have been described as forms of 'ascetic protestantism' and the ideal personality type they attempt to promote—with its hostility to spontaneous enjoyment of life, its tight control over emotional expressiveness, and its systematic ordering of moral behaviour—is, of course, antithetical to the traditional effects of alcohol.[7] It was the ascetic denominations which formed the backbone of the Temperance Movement. In addition, Nonconformism provided the articulate working man with an outlet for his abilities and this, in turn, provided the Temperance Movement with much of its missionary fervour. Of all the dissenting churches, it was perhaps the Methodists who were most militant; whereas in the previous century John Wesley, the founder of Methodism, had seen nothing wrong with a wholesome glass of beer, a hundred years later it would have been a very unusual Methodist who offered a visitor a drink of any kind.

With organized religion to back it, the Temperance Movement attacked the evil of alcohol on every possible occasion and in every possible way. Some of its detestation reached absurd proportions. In the early days and after the victory of total abstinence over moderation had been won, there was a brief dispute over whether medicinal uses of alcohol could be permitted. Again, the more extreme position conquered and, for most temperance supporters, the suggestion of legitimate medical use became an occasion for derisive laughter. Since alcohol was already a product of corruption in the fermentation process, so it was argued, it could not possibly serve any useful purpose at all, although a few more moderate voices reluctantly allowed that it might be necessary in the manufacture of paint and varnish or in the preservation of museum specimens. As for the embarrassingly frequent and apparently approving references to alcohol in the Bible, two dedicated temperance workers devoted their lives to producing a book which proved that what was really intended was non-alcoholic grape juice!

All this propaganda was accompanied by a great deal of pseudoscience and quack medicine, designed to have the effect on people, especially young people, of putting them off alcohol for the rest of their lives. It is no accident that the movement reached the summit of its ambitions during the First World War when a ruthless and well-organized leadership succeeded in playing on pre-existing fears. In Britain, alcohol was blamed for endangering the Allied cause; in the United States, it was implied that drinking was something pro-German and anti-American and that only a total ban could properly prepare the nation for war.

National Prohibition

The great experiment in social engineering which goes by the name of National Prohibition is usually reckoned to have been a disastrous failure, although it did lead to a fall in alcohol consumption, accompanied by a reduction in measures of alcohol-related harm, such as hospital admissions for alcoholism, drunkenness arrests, and liver-cirrhosis mortality. In the opinion of most commentators, however, this is greatly outweighed by the unenforceability of the prohibition law, the widespread lack of respect for the law in general this engendered, and the subsequent upsurge in bootlegging, speakeasies, and gangsterism of all sorts, which has become indelibly associated with the 1920s in the United States. As legal prohibition became increasingly difficult to enforce and as a result, probably, of social changes which made abstinence less attractive to most people, the orientation of the Temperance Movement became increasingly narrow, pessimistic, and vindictive. Desperation over the likelihood of ever succeeding in introducing a truly dry republic led to some unpleasant excesses, such as the denaturing of industrial alcohol with the intended effect of blinding or killing anyone who drank it. In any event, although the Temperance Movement had been highly proficient in getting across its point of view, it is unlikely that this represented the attitudes of the majority of Americans. Once this silent majority began to express its true feelings on the matter, continuing support for Prohibition became politically untenable and repeal an inevitability. The final straw came with the Great Depression when the crisis was taken advantage of by anti-Prohibition forces in the same way as the Great War had been used by those in favour of it. The 21st Amendment to the Constitution, repealing the 18th, was passed by the House of Representatives on 20 February 1933.

The current influence of temperance

Following the repeal of Prohibition, the Temperance Movement has waned considerably in power and the issue of temperance is now remote from everyday considerations. It would be a mistake, however, to imagine that its deeper influence on our thought-patterns and responses to drinking has also disappeared. It is likely that the anxiety, guilt, and ambivalence many people experience in their feelings about alcohol is a direct legacy of temperance campaigning and, moreoever, that the confusion and inconsistency of legal controls over drinking apparent in many parts of the world have been inherited from this source.

Writing in 1967, a leading authority on alcohol studies, Selden D. Bacon, described the continuing impact of the Temperance Movement on contemporary attitudes, action, and research into alcoholism.[8] He pointed out that the movement had led directly to the formation of an anti-movement, representing mainly the interests of the alcohol industry, which responded to the Temperance Movement on a similar level of propaganda, invective, and distortion. This distasteful war between 'wets' and 'drys' had monopolized discussion and fossilized thinking and policy in the area. Even more regrettably, it had resulted in the alienation of the bulk of society from serious consideration of the issues involved, in a repression of humanitarian concern for problem drinkers and in an avoidance of much-needed research and service development by the medical profession, social agencies, universities, and the like. As a consequence, while considerable progress has been made in other problem areas, little had been achieved in the reduction of alcohol-related problems. This Bacon laid squarely at the door of the Temperance Movement and its ossified opposition. Bacon saw signs when he wrote of an improvement in this situation, but subsequent events have proved that the ghost of temperance continues to haunt us.

The nineteenth-century disease concept

In what was more than just a happy coincidence, the notion that 'habitual drunkenness' could be usefully thought of as a disease occurred simultaneously to two medical men on different sides of the Atlantic—Thomas Trotter in Britain and Benjamin Rush in America.

The late eighteenth century was a time propitious for Rush's and Trotter's independent discovery. Ever since the Renaissance, the prestige of science had been growing and it had produced some remarkable achievements. In particular, the foundations of modern medicine had been laid down and the practice of medicine had begun to be associated with some of the prestige it was to acquire in times to come. In this context, it was to be expected that thinkers of originality would attempt to apply the methods and concepts of natural and medical science to human behaviour and the affairs of man; if science had been so successful in explaining the natural world, why should it not be equally successful in explaining the puzzling goings-on of the human race? This decisive moment in the history of thought—the first application of the methods and principles of natural science to the explanation of human behaviour—is sometimes known as the birth of 'positivism' and the creation of the disease theory of alcoholism must be seen as an integral part of it.

It is perhaps difficult now to realize just how revolutionary this proposal was at the time. In the classical view of human nature, which had dominated the seventeenth and most of the eighteenth centuries, man was set apart from the natural world, including all lower forms of life, as being a creature uniquely endowed with reason. The idea that human behaviour was determined by forces outside the individual's control and that it was susceptible to natural, scientific explanations would have been greeted with astonishment by the majority of people; it was implicitly assumed that men always acted freely in accordance with rational principles of self-interest. If some acted in a way contrary to the law of the land or to accepted rules of proper conduct, they were assumed to have done so in full knowledge of the adverse consequences which might follow. The notion that people behaved the way they did, not because they wanted to, but because they could not choose otherwise would have been simply incomprehensible.

From a narrower focus, the newly formulated disease concept can also be seen as part of the emergence of psychiatry as a separate discipline. At about this time, the problem of what to do with the mad was finally being divorced from the response to criminals and vagabonds. The work of the pioneers of psychiatry had resulted in the insane being released from their chains, and the first serious attempts to understand and treat mental illness. The origins of psychiatry have themselves been regarded as part of a more fundamental change in the way social control of deviant behaviour was exerted, with the church and civic authorities being replaced by science and medicine.

The discovery of addiction

Our understanding of the great change in the perception of drunkenness at the end of the eighteenth century has been greatly assisted by the work of sociologist Harry Levine and his essay *The discovery of addiction*.[9] Following Levine, it will again be useful to contrast the first disease concept with what went before. Although it was never regarded as a serious threat to the social order, we have seen that drunkenness was sometimes perceived as a problem in the classical view—a problem of individuals requiring punishment or, occasionally, of special segments of the population requiring legislation. But the crucial point is that, although drunkenness may have been seen as a problem, it was not seen as problematic, in the sense of 'requiring an explanation'. In Levine's words, 'drunkenness was a choice, albeit a sinful one, which some individuals made'; there was nothing here to explain, only to correct.

While some observations on the addicting nature of distilled spirits were being made by his contemporaries, it is in the work of Dr Benjamin Rush of Philadelphia (1745–1813) that we find the first, fully formed conception of alcohol addiction as a disease. Rush was the foremost medical practitioner of his time. Apart from being the father of American psychiatry, of the American public health movement, and, as we shall see, of the American Temperance Movement, Rush was a friend of Benjamin Franklin and Thomas Jefferson and a signer of the Declaration of Independence. A staunch ideological republican, he shared with his friends the vision of America as a stable, just and free society composed of virtuous and dutiful citizens. Anything which threatened this vision was sure to provoke Rush's tireless opposition. It was in this mood that he warned his fellow countrymen that excessive spirit drinking, far from being the beneficial activity it was commonly reputed, in fact endangered the very future of the nascent republic.

Rush's classic work was published in about 1785 under the forbidding title of *An inquiry into the effects of ardent spirits upon the human body and mind with an account of the means of preventing and of the remedies for curing them.*[10]

As a scientific document, the *Inquiry* is a curious mixture of pious moralizing and enlightened observation; its main interest lies in the clear statement that a form of alcohol is a causal agent of addiction and the unambiguous designation of addiction to spirits as a 'disease of the will'. Rush wrote that once an appetite or 'craving' for spirits had developed, the victim was powerless to resist and could no more stop an impulse to drink than a convulsive movement of a hand or foot. In these cases, drunkenness had ceased to be a vice or personal weakness since the drunkard had completely lost control over his drinking; rather, the drunkard was himself controlled by drink. Rush provided a vivid illustration of this helplessness in the words of one of his patients who said that, even if cannon balls were discharging between a keg of rum and the place he was standing, he would be unable to prevent himself from trying to get a drink. It may have been reports of irresistible cravings for alcohol such as this which influenced Rush, but there can be little doubt that his formulation in turn encouraged many excessive drinkers to see their problem in this way. Indeed, it is a surprising fact that alcohol began to be seen as an addicting substance some seventy years before opium was so regarded.

In a later edition of the *Inquiry*, Rush tried to capture the progressive nature of the disease in the Moral and Physical Thermometer shown in Fig. 2.2. In the upper part of the thermometer, beverages ranging from cider to strong beer are linked, if taken in moderation, with cheerfulness, strength, and nourishment. However, in the lower half, the use of increasingly strong spirits is associated with a gradual deterioration in physical health and social behaviour and an increase in the adverse consequences of drinking. In the next chapter we will compare Rush's thermometer with a more recent graphic depiction of the course of alcohol addiction (see p. 82).

Having diagnosed the disease, Rush then prescribed the treatment. In his view, there was only one possible cure for the addiction—complete abstinence from spirits. Thus, although he was not opposed to all forms of alcohol and, indeed, recommended cider, malt liquor, and wine as substitutes for spirits, Rush did anticipate the later hardline temperance

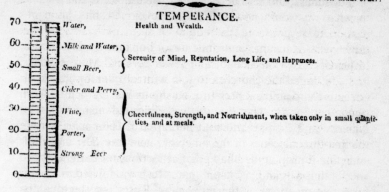

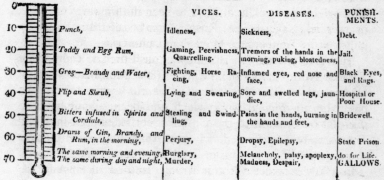

Fig. 2.2. Rush's Moral Thermometer (reproduced by permission of the British Library).

position by ruling out the possibility of moderate use of the addicting substance. As for the means to be employed in achieving the desired goal of abstinence, the remedies advocated by Rush seem rather harsh and include frights, beatings, whipping, threats, and cold showers.

Rush was also concerned with wider issues than the treatment of individual drunkards and devoted some space in the *Inquiry* to what we would now call social policy. Despite his compassionate response to the addict in need of treatment, he recommended harsher punishments for intoxication, in addition to heavier taxes on alcohol and fewer taverns. More significantly, he urged the churches to join with him in a campaign of education and political pressure. Rush's appeal certainly fell on sympathetic ears. The *Inquiry* was reprinted many times in America and Europe and sold perhaps 200 000 copies during the first three decades of the nineteenth century. The model of addiction it propounded had great attractions for early temperance enthusiasts and, as an old man, Rush was invited to attend formative meetings of the movement. Later, the teetotallers continued to acclaim Rush as their inspiration and patron, although his benevolent attitude to beer and wine was tactfully overlooked. Nevertheless, Rush is rightly called the founding father of the Temperance Movement in America.

Rush's counterpart in Britain was the Edinburgh physician Thomas Trotter (1760–1832) who, in 1804, published *An essay, medical, philosophical and chemical on drunkenness*.[11] This was based on an original work which had been submitted as a doctoral dissertation to the University of Edinburgh in 1788. The *Essay* became immediately popular, rapidly went through several editions, and was translated into Swedish and German.

In a key sentence, Trotter wrote: 'In medical language, I consider drunkenness to be a disease, produced by a remote cause, and giving birth to actions and movements in the living body, that disorder the functions of health.' In further discussion of the nature of this disease, Trotter insisted that it was not only a bodily infirmity which needed correction, but that 'the habit of drunkenness is a disease of the mind'.

Unlike Rush, Trotter recommended total abstinence from all alcoholic beverages as treatment for the disease. His suggestions for arriving at abstinence were somewhat subtler than

Rush's and he realized the futility of simply moralizing and preaching to the patient. He was aware of the importance of the environment and recognized that it was not good enough simply to return the patient to the same conditions he had come from.

Disease theory and the Temperance Movement

The main contribution of Rush and Trotter to the ideology of the Temperance Movement was in providing it with a way of understanding and publicizing to the world at large the evil effects of alcohol and the dangers attendant on anyone's drinking. Describing habitual drunkenness as a disease meant that there could be little rational opposition to its control and ultimate eradication. According to Levine, later temperance workers looked hard for public declarations by former inebriates of the addictive properties of alcohol, but the first they were able to find was written in 1795. Following the spread of Rush's and Trotter's ideas, however, it soon became commonplace to hear drunkards attest to the involuntary and, hence, disease-like nature of their drinking, and the vocabulary of irresistible craving and compulsion was increasingly used to describe the drunkard's inner experience. This vocabulary of addiction became the major organizing theme of temperance thought and literature and, indeed, the deranged and pathetic victim of addiction to alcohol was to become a stereotype of that unique nineteenth-century literary form, the Victorian novel. What is essential to grasp, however, is the similarity between these nineteenth-century descriptions of the drunkard's experience of alcohol and modern, disease-theory accounts of the alcoholic's 'loss of control' over drinking. This similarity is illustrated by the following portrayal of the drunkard's predicament from a temperance writer in 1833:

In their sober intervals they reason justly, of their own situation and its danger; they know that for them, there can be no temperate

drinking: They resolve to abstain altogether, and thus avoid temptation they are too weak to resist. By degrees they grow confident, and
secure in their own strength, and . . . they taste a little wine. From
that moment the nicely adjusted balance of self control is deranged,
the demon returns in power, reason is cast out, and the man is
destroyed.[9]

Ignoring unimportant differences in manner of expression,
this is virtually identical in meaning to accounts contained in
any Alcoholics Anonymous (AA) pamphlet describing the
devastating effects of 'the first drink' on a previous abstinent
alcoholic. It shows that the AA concepts of craving and loss of
control, concepts which lie at the heart of the AA understanding of alcoholism and which have been taken over in large part
in psychiatric disease models of alcoholism, are inherited from
nineteenth-century temperance ideology.

The similarity between old and new versions of loss of control
becomes especially obvious when one reads contemporary
accounts of the arguments between the moderationists and
teetotallers which occurred mainly in the 1830s before the latter
took control of the Temperance Movement. The scorn and
ridicule poured forth by the teetotallers on the whole idea of
moderation, and the personal testimonies given to the ineffec-
tiveness of attempts to restrain drinking and to the inevitability
of relapse, appear almost exactly the same as modern
objections by members of AA to the possibility of controlled
drinking by alcoholics (see p. 90).

But there is, of course, one crucial difference here. The
nineteenth-century teetotallers decried the possibility of control
over drinking by *anybody*, not merely a special kind of person
called an alcoholic. For them, the essence of alcohol addiction
lay in the addicting nature of alcohol itself, not in the special
vulnerability of a few unfortunate individuals, as in the AA
theory. It is as though AA borrowed the temperance view of
addiction but simply cut off its range of application, restricting
it from everybody to a predetermined few.

There are further similarities between the early temperance disease concept and its modern, AA counterpart. For example, many temperance writers stated that addiction to alcohol was often inherited, thus anticipating the AA claim that a special vulnerability to alcohol was derived from an inborn, constitutional deficit. It must be emphasized that the temperance writers believed in an odd sort of inheritance, typically subscribing to a version of the inheritance of acquired characteristics in which the sins of the parents could be visited on the offspring. Moreover, the Victorian fascination with heredity does not detract from the fact that addiction was regarded as a potential danger to anybody who drank.

More important is the similarity between earlier and later attitudes to 'inebriates', as they began to be called in the latter part of the nineteenth century. Contrary to much that has been written, temperance supporters were not hostile to the individual inebriate but forgiving and compassionate. There was no opposition from the main body of the nineteenth-century Temperance Movement to the provision of treatment for alcoholism and, indeed, help for inebriates formed a major part of temperance activity. It was for the moderate drinker that contempt was especially reserved, on the ground that, in appearing to believe that moderate drinking was harmless, such a person represented a dangerous precedent to those whose lives could well be ruined by alcohol. Above this, anger was directed against the purveyors of alcohol and a society which allowed this practice to continue. Conversely, opposition to the Temperance Movement did not come from adherents of the disease position but from the opposite point of view, held by those who retained the classical 'free-will' philosophy which insists that alcoholics are still responsible for their actions (and, of course, there are still plenty who hold this view today). Antagonism between temperance supporters and those concerned with providing treatment and help for alcoholics arose later, mainly in the present century, because of a change of

direction in temperance aims. It is certainly not true that the
concept of alcohol addiction as a disease weakened the basis of
temperance ideology; instead it strengthened it.

The Washingtonians

Nowhere is this more plainly demonstrated than in the activities
of the Washingtonians, a fraternal self-help association for
reformed drunkards which enjoyed a tremendous wave of
popularity during the 1840s.[12] Started by a group of six topers
in Baltimore in 1840 and named in honour of the first president,
the Washingtonian movement rapidly swept the United States
and by 1843 could number millions among its adherents. From
its ranks arose the greatest of all temperance orators, John B.
Gough. Chiefly because of Gough's visits, a similar wave of
popularity occurred in Britain.

It is the organizational form adopted by the Washingtonians
which is of chief interest here. Meetings of local groups were
based closely on the model of the revivalist religious meetings
which were popular at the time and typically consisted of a
series of autobiographical 'confessions' regarding the individual's former battles with alcohol, culminating in the trium-
phant victory of complete abstinence over the unnatural
appetite for drink. The language and concepts employed in
these recitals were taken straight from the disease theory of
addiction and were little more than amplified personal testimonies to the overwhelming power of 'loss of control' and the
indispensability of total abstinence. Apart from such meetings,
the principal activity of members was the attempt to reclaim
the lives of other drunkards and gather new recruits. This was
regarded as essential, not only on humanitarian grounds, but
as an aid to the member's own salvation. Besides the direct
support for sobriety given in this way, the Washingtonians
provided the opportunity for an active social life without
alcohol and a sense of brotherhood and collective identity.

Great emphasis was placed on the integrity of the family and auxiliary groups were formed for wives and for children. The reader with any acquaintance with the characteristics of AA will immediately recognize in this brief description the main features of the modern self-help organization (see pp. 49–54). As a leading historian of British temperance has observed, 'the early teetotal movement pioneered many of the remedies which have since been rediscovered by Alcoholics Anonymous'.[13]

There are also important differences. Although the Washingtonians were sometimes highly secretive and held closed meetings from which non-members were excluded, there was not the same emphasis on anonymity which is an important feature of AA. The Washingtonians included among their ranks many moderate or even light drinkers, as well as many who would today be labelled alcoholics, and were dedicated to the temperance ideal of abstinence for all, unlike AA who admit only self-confessed alcoholics to the fellowship and concern themselves with nobody's drinking but their own. Finally, unlike the Washingtonians, AA claims not to get involved in political issues and has far more sophisticated methods of encouraging and sustaining abstinence than its nineteenth-century counterpart.

Nevertheless, on a more fundamental level, both organizations owe their existence to a deeply embedded cultural role found in societies, like those of the United States and Britain, which have been influenced by evangelical Protestantism. This has been called 'the repentant role'; it is characterized by repeated confessions of guilt and avowals of repentance and its function is to provide the 'reformed sinner' with a special but acceptable place in the social order.[14] Although the Washingtonians were ostensibly a secular organization and were opposed by clerical elements in the Temperance Movement, they shared with AA an orientation which is best described as quasi-religious.

While the particular phenomenon known as the Washington-ian Movement underwent a dramatic decline at the end of the 1840s, many similar temperance societies, dedicated to the goal of mutual aid for inebriates and using the Washingtonians as their prototype, continued to flourish throughout the century in both America and Britain. There can be little doubt that these societies were the developmental forerunners of AA and that the nineteenth-century theory of alcohol addiction as disease was conveyed to it in this way.

The inebriate asylums

The connections between the medical profession and the Temperance Movement did not end with the work of Rush and Trotter. In both America and Britain, medical men were active in the movement, especially by giving lectures and writing tracts. In the early years, however, medical practitioners had not yet attempted to carve out for themselves a unique role in the response to alcohol addiction. This took place in the second half of the century with the advent of special asylums for the treatment of inebriates. Such an idea was originally proposed by Rush, who recommended the setting-up of a 'sober house' where drunkards could receive specialized medical attention. It was also endorsed in 1838 by Samuel B. Woodward, the superintendent of one of the first mental asylums and the leading 'alienist', as psychiatrists were then called, of his day. Woodward also emphatically endorsed a disease model of habitual drunkenness with abstinence as its only cure and advocated asylums as places where abstinence could be enforced. The first inebriate asylum was established in 1841 in Boston, Massachusetts, during the wave of popularity for the Washingtonians, but had to close through lack of funds when this enthusiasm died away. However, it reopened in 1857 and began to receive the support of important temperance organizations. From that point on, the number of asylums in America steadily increased until it reached over fifty by the end of the century.

In 1872, the superintendents of these inebriate asylums banded together to form the Association for the Study of Inebriety. Like most bodies of its kind, the primary concern of the Association was with the professional interests of its members and it sought to establish both alcohol and narcotic addiction as recognized neuroses or psychoses and areas of legitimate medical specialization. It was less interested in communicating with the general public than in its status and credentials among the remainder of the medical profession and campaigned with some vigour for the removal of the inebriate from the prison system, in favour of compulsory incarceration in treatment programmes. To further these ends, the Association began to publish, in 1876, the *Journal of Inebriety*, which included papers dealing mainly with the disease nature of alcohol addiction. In its early days, the quality of papers in the journal was high and many of the concepts which were later to be utilized by Jellinek in his celebrated disease theory, including the concepts of increased tolerance to alcohol, adaptation of nervous tissue, loss of control, and craving, were known to the early writers on inebriety and fairly accurately described.[15]

The Association grew rapidly and received considerable support and encouragement from the Temperance Movement. It is true that a few temperance clergymen were hostile at first and that some ordinary temperance workers were too imbued with the nineteenth-century idea of self-help to accept the principle of medical treatment. On the whole, however, temperance leaders were favourable to inebriate asylums.

The Association was less successful in its appeal to the medical fraternity. This may have been because of some lack of scientific rigour in its pronouncements and, after about 1900, some of the disease formulations contained in the *Journal of Inebriety* were vague and ill-defined. Possibly as a result, it was forced to cease publication in 1914. But it is unlikely that

medical treatment of inebriety was rejected by the majority of
the profession on purely scientific grounds. Despite the efforts
of temperance supporters, working with inebriates still carried
considerable stigma and was accorded a low status by doctors
even within the poorly regarded speciality of psychiatry. This is
still true today.

Keeley's cure

Among the general public, inebriate asylums met with a mixed
reception and were far less successful than other remedies that
were proposed for inebriety. In 1880, Dr Leslie Keeley
announced the discovery of a remarkable cure for intemperance, established an institute in his home town of Dwight,
Illinois, and began receiving patients.[4] Keeley's cure consisted
of 'double chloride of gold', a compound unknown to pharmacology but apparently imbued with secret powers of relief for
the drunkard when injected intravenously or taken orally.
Owing to press coverage, this cure suddenly became enormously popular and drunkards flocked to a proliferating
number of Keeley Institutes in the hope of benefiting from it.
Thousands of former patients were turned into missionaries to
spread the gospel and were formed into leagues which held
conventions where the wondrous properties of the cure were
extolled by hired lecturers. There was even a grand exposition
in Chicago in 1893 at which hundreds of Keeley graduates
paraded in honour of their saviour. Keeley died a rich man in
1900 but enthusiasm for the cure did not long outlive its
inventor, probably owing to debunking by other medical men
and charges of quackery. But before its demise, some 400 000
people had taken the Keeley cure.

This tale should not be regarded as merely a historical
curiosity because two important lessons may be learned from
it. First, the popularity of a cure for alcoholism may bear little
relationship to its scientific validity. Secondly, the story of

Keeley's cure shows how powerful is the attraction of a simple and undemanding solution to the problem of alcoholism. In subsequent chapters, we shall discover that this hunger for a simple and easily understood solution has by no means disappeared.

Alcoholism and the growth of psychiatry

While various suggested treatments were experiencing changes of fortune, the disease of inebriety was finding its way with increasing regularity into the medical textbooks and was gradually becoming a recognized part of academic psychiatry, if not of psychiatric practice. The development of medical and psychiatric understanding of alcoholism has been fully described elsewhere and will not be repeated here in any detail.[16] Only the main advances in terminology and classification will be briefly mentioned.

Following Rush and Trotter, attempts to place the study of habitual drunkenness on a scientific footing passed to the continent of Europe. The great French psychiatrist Esquirol, in 1838, classified it as a 'monomania' whose principal characteristic was 'an irresistible tendency towards fermented beverages' and compared its driven, deterministic quality with the obsessional-compulsive neurosis he was the first to describe. A similar formulation was advanced by the German, Brühl-Cramer, who had no doubt that he was dealing with a disease of physical origins. The term 'dipsomania', which later came to have the special meaning of periodic bouts of drunkenness, was derived from Brühl-Cramer's work.

The term 'chronic alcoholism' was coined by a Swedish professor of medicine, Magnus Huss, in 1852, although it did not come into general use until the twentieth century. Huss provided the first systematic description of an alcoholism syndrome, thoroughly backed up by extensive illustrations from case studies, and anticipated several important advances in the understanding of the physical damage caused by excessive drinking. Finally, in 1901, the great classifier of psychiatry, Emil Kraepelin, used the term 'alcohol addiction' to describe those who were unable to give up the harmful use of alcohol, and other authors of this time saw alcoholism as an expression of an underlying circular psychosis. In this way, alcoholism was related to some of the major categories of psychiatric thought.

It might be imagined that these developments accelerated the acceptance of the disease concept and treatment for alcoholics outside the medical profession. In fact, at around the end of the nineteenth century, just the opposite happened. One of the reasons for this was a change of direction by the Temperance Movement from a largely educational and reformist approach to a wholehearted and exclusive devotion to the drive for legal prohibition. The older fraternal organizations, aiming at mutual aid for the victims of alcohol, declined markedly. One consequence was that interest in the concept of addiction was significantly reduced and was more or less dropped from temperance ideology. (The concept of addiction as such did survive but was transfered from alcohol to opium, marijuana, and other drugs.) Correspondingly, interest in the individual alcoholic was replaced by a single-minded concentration on the evils of the liquor industry and the saloon. Drunkards ceased to be objects of pity and concern for temperance workers and became regarded merely as pests.

In America, this tendency became even more pronounced during Prohibition, when it was assumed that any problems caused by alcohol would simply go away. This neglect virtually eliminated any interest shown in the subject by the medical profession and allied occupations and the Association for the Study of Inebriety was finally dissolved in the mid-1920s. By the time Prohibition was repealed in 1933, it was almost as though the concept of alcohol addiction as a disease had never existed.

The disease concept rediscovered

Following repeal, any attempt by government to interfere in the drinking habits of Americans would have been politically disastrous. Consequently, the new Roosevelt administration assiduously avoided the issue and no coherent government policy was formulated. The task of responding to alcohol problems, which began to increase directly Prohibition was lifted, was left to private interests. Moreover, it was almost inevitable that any private attempt to revive the nation's concern over drinking should be accompanied by the aura and prestige of science. People were tired of the emotionalism and extremity of views which characterized the battle between wets and drys and were only likely to respond to an approach which seemed to convey the objectivity and rationality of science.

The Alcoholism Movement

It was against this background that, in 1937, a group of physicians and natural scientists founded the Research Council on Problems of Alcohol, with the object of raising funds for medical research into alcoholism.[17] Although it was largely unsuccessful in this, the prominence of some of its members did lend the Council an air of scholarly respectability and enabled it to obtain a grant for a review of the literature on the biological effects of alcohol on humans. A certain Dr E. M. Jellinek, who was then doing research into the biochemistry of schizophrenia, was persuaded to administer the project. Eventually, a member of the Council, Howard W. Haggard, who was director of the Laboratory of Applied Physiology at Yale University and who had been supervising research into alcohol metabolism, invited Jellinek to Yale to head a multidisciplinary research centre.

The Yale Center for Alcohol Studies was to be immensely influential in the development of a treatment response to alcohol problems in America and, by extension, in the rest of the world. In 1940, Haggard started the *Quarterly Journal of Studies on Alcohol* (changed to the monthly *Journal of Studies on Alcohol* in 1975), which remains the most prestigious journal in the field. Apart from the outpouring of research which found its way into the journal, the Yale Center began in 1943 to hold Summer Schools attended by representatives of AA, the Temperance Movement, and the liquor industry, as well as many citizens from all over the country involved in the formation of alcohol policies in their local communities. According to Robert Straus,[18] who was involved at the time, the organizers of the Summer Schools adopted a model of alcoholism as a disease in a deliberate and calculated strategy to combat stigma and prejudice, encourage alcoholics and their families to seek help, and improve the image of the alcoholic among the helping professions. It was also intended to distance the alcoholic from other groups with more obviously spoiled social identities, such as drug addicts and the mentally ill. The Summer Schools did much to disseminate the idea that alcoholism was a disease and that the alcoholic could be treated.

The Yale Center also supported the formation of the National Committee for Education on Alcoholism. This was set up by three women, including Mrs Marty Mann, one of the first female members of AA. In 1950, it changed its name to the National Council on Alcoholism; it continues to be the leading voluntary organization in the field in America, supplementing the work of AA and presenting a strictly disease-oriented view. Yet another development stimulated by the Center was the Yale Plan clinics which offered out-patient rehabilitation at minimal cost and provided a prototype for many later treatment facilities. In 1962, owing apparently to a lack of enthusiasm for the publicity its activities had aroused at Yale, the Center moved to Rutgers University, where it has remained ever since. The Rutgers Center was financed largely from money provided by the recovering alcoholic millionaire and member of AA, Brinkley Smithers.

Although this brief description of the activities of the Yale Center will have made clear its intimate link with the Fellowship of AA, a detailed consideration of the remarkable rise of AA will be left to a later section of this chapter. At this point, however, we shall merely mention that the triumvirate of the Yale Center for Alcohol Studies, AA, and the National Council on Alcoholism is often referred to collectively as 'the Alcoholism Movement'.

The new approach

So dormant had the disease theory become during the first part of the twentieth century that the rediscovery of this idea was labelled 'the new approach to alcoholism'. Despite the scientific aura with which it was surrounded, this new approach was motivated very much by practical rather than scientific considerations. As Robin Room, a leading thinker in the field, has commented,[19] 'the promulgation of disease concepts of alcoholism has been brought about essentially as a means of getting a better deal for the "alcoholic" rather than as a logical consequence of scholarly work and scientific discoveries'. The rediscovered disease concept was left deliberately vague so as not to offend powerful lobbies and to unite interested bodies with diverse aims and views. The main object of the exercise was to keep the alcoholic out of gaol and get him or her into treatment, not to satisfy the academic standards of scientific research and theory.

Another feature of the rediscovered disease concept requires careful consideration. In the years after repeal, no attempt to promote sympathy for the alcoholic was likely to attract general support if it inveighed against the dangers of alcohol itself. The entire notion of temperance had been thoroughly discredited by the experience of Prohibition and anything that smacked of an 'anti-alcohol' sentiment was certain to generate suspicion and hostility in many quarters. Not only that, but historical changes had already taken place which had the effect of rendering teetotalism a bizarre and irrelevant posture. The ninenteenth-century ideals of frugality, hard work, and self-control had been largely replaced by an emphasis on consumerism, leisure, and the unfettered pursuit of pleasure, and there was an increasing desire to be free of the old restraints on personal conduct imposed by the family, church, and local neighbourhood. Alcohol was a symbol of this new freedom (as was, incidentally, the cigarette). In the same way that abstinence had displayed adherence to the old ascetic values, drinking symbolized their rejection.

For all these reasons, it was essential for the new approach to avoid appearing to criticize alcohol and to recognize the apparent fact that millions of people drank without problems. Therefore, so the logic went, alcohol itself could not be the cause of alcoholism. In this way, of course, the new account of addiction represented a sharp contrast to the nineteenth-century version, endorsed by the Temperance Movement, which maintained that the agent of the addiction was alcohol and that all drinkers were liable to become addicted. It was now said that only a few, predisposed individuals were susceptible to the disease; drinking by the great majority did not lead on to addiction. In the mood of the times, this was the only kind of disease concept which stood any chance of getting off the ground.

The relationship between the Alcoholism Movement and temperance supporters was at first very cordial. Several prominent temperance leaders hailed the disease concept and treatment for alcoholics as an important and welcome advance. It was only later, after temperance workers had come to perceive common interests between the Alcoholism Movement and their natural enemy, the liquor industry, that the two movements quarrelled. Indeed, there is evidence that the industry commissioned two private studies of the implications of the disease concept, both of which reported that it had nothing to fear.[20] The reasons were that the new disease theory did not see anything wrong with drinking for the great majority of people and that the only consumers liable to be harmed by the industry's products were predestined to be so. Thus the disease concept provided a very convenient exculpation of the activities of the liquor industry which is still appealed to by its spokesmen today.

It must be stressed again, however, that there is a strong continuity between the rediscovered disease concept and its predecessor advanced by Rush and Trotter, developed by the managers of inebriate asylums, and rooted in temperance ideology. The essential similarity lay in the concept of addiction itself and in the description of the addictive process through the use of terms like loss of control and craving.

Success of the disease theory

According to several formal criteria, the struggle to have alcoholism recognized as a disease appears to have succeeded. Although the American Medical Association had first given approval for the inclusion of alcoholism in its official nomenclature of diseases as early as 1933, this had had little effect. In 1956, the Association threw its weight behind the disease concept and stated that the medical profession in general 'recognizes this syndrome of alcoholism as an illness which justifiably should have the attention of physicians'. Equally important in the 1950s was the formal acknowledgement of the disease concept given by special subcommittees of the World Health Organization, of which Jellinek was always a prominent member.

The 1960s witnessed several significant developments. In

1966, two Federal Appeal Court decisions supported the disease concept (see pp. 162–4) and there was a rapid growth in the number of companies with programmes for the treatment of alcoholism among their employees. In addition, the 1960s saw an increasing interest by federal government in the national response to alcoholism, thus overcoming the reluctance to get involved which was an aftermath of the Prohibition experience. A moving force behind these developments was Senator Harold Hughes, a confessed recovering alcoholic and AA affiliate.

These advances culminated in the signature by President Nixon on the last day of 1970 of a bill establishing the National Institute on Alcohol Abuse and Alcoholism (NIAAA). The remit of this agency was to set up treatment programmes throughout the United States and to encourage the development of special programmes for prevention, training, and research. In the first year of its existence, NIAAA's appropriation was thirteen million dollars but this had risen to 175 million dollars before the end of the decade. NIAAA and the growth of private treatment facilities in America has resulted in a veritable 'alcoholism industry', employing large numbers of people with vested interests in its continuation.

The progress of events in Britain has followed a similar course, although lagging behind those in the United States. The first London branch of AA was founded in 1948 but this aroused very little interest among the medical profession. Indeed, when in 1951 a consultant psychiatrist asked for funds to attend a WHO conference on alcoholism, his application was rejected on the ground that 'there was no alcoholism in England and Wales'.[21]

Following this low ebb, the problem of alcoholism became gradually acknowledged and the disease concept was officially endorsed by the British Medical Association. The model of alcoholism adopted in the NHS has been based on the pioneering work of Dr Max Glatt and the treatment centre he established at Warlingham Park, Middlesex, in the early 1950s.

This consisted of a specialized and self-contained in-patient unit for alcoholics with an emphasis on group therapy and close links with AA. A memorandum issued by the Ministry of Health in 1962 recommended the setting-up of at least one such unit in every region of Britain and this was slowly but surely accomplished over the next twenty years.

As will be obvious, this model of treatment is in direct line of descent from the inebriate asylums of the last century. Moreover, in adopting such a model, the treatment response to alcoholism ran counter to the swing towards out-patient treatment and community involvement which took place in the mainstream of psychiatry. More recently however, and following a circular from the Department of Health and Social Security in 1973 entitled *Community services for alcoholics*,[22] there has been a growing impetus towards a community approach to problem drinking in Britain.

Alcoholics Anonymous

On 10 June 1935, in Akron, Ohio, there occurred a chance meeting between William Wilson, a stockbroker from New York visiting Akron on business, and Robert Holbrook Smith, a local physician. Both men were alcoholics and had begun having problems with their drinking during Prohibition. At one time, Wilson seemed to have had a brilliant career on Wall Street waiting for him until his drinking had taken over. The previous December, he had been in hospital suffering from severe withdrawal symptoms when, in the depths of his despair, he had experienced a moment of spiritual enlightenment. From that moment on he had not touched alcohol, and had become convinced that he could keep sober by admitting to himself and to God that he was powerless to control his drinking and helping others to reach the same insight. Unfortunately, he had failed to persuade any other alcoholics of this wisdom and now, under pressure from yet another failed business venture in Akron, he feared a relapse.

The physician too had recognized for some time that he was incapable of controlling his drinking but had been unable to do anything about it. As the men talked, they began to share their disastrous experiences and agreed that they were the victims of a disease. The broker came to realize he had found the fellow sufferer he was looking for. They resolved to help one another stay sober and to spread their message of hope to other alcoholics. The two parties to this chance encounter subsequently became internationally famous—not by their own names but under the abbreviated names of 'Bill W.' and 'Dr Bob', the founders of Alcoholics Anonymous.[23]

Besides their drinking problems, Wilson and Smith shared an involvement with the Oxford Group, a popular and aggressively evangelistic religious movement of the 1930s and forerunner of Moral Rearmament. After his religious experience in hospital, Bill W. had reinforced his 'spiritual awakening' and kept away from drink by attending Oxford Group meetings in New York and abiding by their precepts. Dr Bob had also had extensive involvement with the movement in Akron. For the first two years of AA's existence, the co-founders were able to maintain active membership of both groups until important differences in goals became apparent.

This early connection points to the religious and spiritual character of AA which was present from the beginning. AA has always denied it is a religious organization but would readily admit to the spiritual quality of its message. *The Twelve Steps*, the body of principles used for recovery which forms the backbone of AA teaching, contains only two mentions of alcohol but five references to God. The first and all-important Step is: 'We admitted we were powerless over alcohol—that our lives had become unmanageable.' In one sense, this is a restatement of 'loss of control', the fundamental principle of the nineteenth-century concept of addiction. However, the abdication of personal responsibility, implied by the First Step, leads to the more obviously religious content of the Second and Third: 2. '[We] came to believe that a Power greater than ourselves could restore us to sanity'; and 3. '[We] made a decision to turn our will and our lives over to the care of God as we understood Him.'

The last phrase stresses that AA is completely non-denominational but also means that 'God' need not be interpreted even in the sense of an Almighty. Some AA affiliates, possibly of an atheistic or agnostic persuasion, see the 'Power greater than ourselves' as the indispensable mutual aid provided by the Fellowship itself. Whatever interpretation is made, however, there can be little doubt that the principles and activites of AA have much in common with those of religious groups. The remainder of the Twelve Steps are concerned with the issues of faith, resignation, confession of wrongs, the need for spiritual awakening, and the duty of spreading this awakening to others. Bill W. himself affirmed that the emphasis on guilt, confession, and repentance in AA, as well as the idea of group meetings, came straight from the Oxford Groups. Although the only qualification for membership is a genuine desire to stop drinking, AA clearly derives much of its character from evangelical Protestantism.

This does not mean that the tone of AA activity itself is evangelistic. In the days before his meeting with Dr Bob, Bill W. had been told by his own physician, Dr William D. Silkworth, that his failure to convert other alcoholics had been due to too much preaching. Silkworth, whose importance to the newly born movement is such that he is referred to as 'the patron saint of AA', advised a more 'scientific' approach. He even suggested the specific content of this approach in the idea that alcoholics had a 'mental obsession' with drinking because they possessed a 'physical allergy' to alcohol. One of the key factors in the success of AA was its ability to convey an essentially moral and religious message in an ostensibly scientific package.

The growth of AA

Following their momentous encounter, Bill W. and Dr Bob began establishing local AA groups, which held regular meetings and were open to anyone who was interested. At first, progress was slow. When in 1939 the *Big Book* of AA was published,[24] setting out the basic principles of AA philosophy, there were fewer than 100 members. However, the *Big Book* attracted publicity and several newspapers and national magazines ran articles on AA, including a famous piece in the *Saturday Evening Post*. The result was astonishing. As with the Washingtonians a century earlier, membership expanded at an explosive rate. By the end of 1940, there were 2000 members and by 1941 this had become 8000. Praise was heaped on AA from all quarters and leading members were invited to the White House to meet the President. By 1957, just after the American Medical Association had recognized alcoholism as a disease, the estimated membership was 200 000. AA's promotional activities were considerably helped by the respectability given to it by the scholars and scientists of the Yale Center and by the indefatigable efforts of the National Council on Alcoholism. But even without this assistance, it is clear that the early growth of AA is a typically American success story.

AA has continued to expand, although not at such an exponential rate. Since no records are kept of local group meetings, it is impossible to arrive at precise figures and it has been argued that AA estimates of membership are exaggerated because they do not take account of the tendency for members to belong to more than one group. However, the Fellowship has calculated that in 1985 there were over one and a half million AA members in 40 000 groups in over 100 countries throughout the world. Membership is still growing, at a rate in

Britain calculated at 15 per cent per annum. The latest area of the world for rapid expansion is Latin America.

AA's numerical success has been matched by its ability to promote itself as the best chance alcoholics have for recovery. For example, in a survey conducted in the United States in 1972, the majority of physicians interviewed thought that referral to AA was the wisest professional strategy to adopt on behalf of alcoholic patients.[25] Many hospital rehabilitation programmes consist of little more than a formalized version of AA principles or make attendance at AA meetings a compulsory part of treatment. Among the general public too, a community survey in London showed that nearly 50 per cent of those interviewed believed AA to be the best source of help for people with drinking problems.[26]

Soon after its inception, AA began to evolve its characteristic structure—or rather lack of structure, since AA is at pains to insist that it is not an organization in the normal sense of the word, preferring the term 'Fellowship'. The power in AA is deliberately designed to lie at the periphery, in the local groups, with only a minimal centralized administration. Moreover, no fees or dues are received and AA is entirely self-financing, relying on voluntary donations and collections from members, or 'affiliates' as they prefer to be called. Outside contributions of as much as £5000 have been declined. Equally important is the principle of anonymity which, besides its protective function, is regarded as a disincentive to the seeking of publicity or other self-aggrandizement by individual members. AA maintains that it engages in no enterprises or outside activities, does not express a view on any issues or controversies, and does not endorse any political or other body of opinion, such as temperance. In 1950 these principles were codified as *The Twelve Traditions of Alcoholics Anonymous*, to match *The Twelve Steps*, and have been more or less faithfully adhered to since. These characteristics of AA have been the inspiration for other self-help groups dealing with several different kinds of problem.

Of the many different ways in which commentators have tried to capture the essence of AA, one would probably meet with general agreement. This is that AA offers not so much a method of recovering from alcoholism as a completely new way of life. The close-knit values and integrated activities of AA represent a self-contained sub-culture offering the individual a ready-made, fresh identity with which to confront life without alcohol. The idea that AA substitutes dependence on the Fellowship for dependence on drink is not one most affiliates would reject. This all-embracing aspect of AA is accentuated by the existence of thriving sister organizations: Al-Anon, for the spouses and friends of alcoholics, and Al-Ateen, for their teenage children.

Who goes to AA?

An important question concerns the possibility that AA may be attractive to certain types of problem drinker and not to others, despite the fact that it is open to anyone who wishes to stop drinking. The existence of selective factors is suggested by findings that only 10 to 20 per cent of problem drinkers who are referred to AA ever attend meetings regularly; the rest either attend on an irregular basis or drop out completely.

There is evidence that successful AA affiliates differ from other kinds of problem drinker in personality. AA appears more suited to those of a more authoritarian type of personality who see things in black-and-white terms and who enjoy belonging to groups that provide a clear structure for understanding the world. From a different angle, there is ample evidence that AA members have a strong need for affiliation and show a greater than average dependence on other human beings; they are more 'clubbable' kinds of people than those who do not take to AA.[27] Although there is probably a great deal of variation across different AA groups in different cultures, there is also evidence that AA appeals more to white, socially stable, middle- to upper-class problem drinkers, whose lives are comparatively satisfying when free from alcohol problems, than to lower-class individuals or those from other ethnic backgrounds. As might be expected from what has already been established about AA, the Fellowship attracts people who are either overtly religious in orientation or, at least, have existential concerns related to religious belief. Such AA members are frequently found to be somewhat conservative in their social attitudes and to be more than usually concerned with 'respectability'. Individuals who do not possess this kind of personality or who do not share the necessary religious or social attitudes are likely to be put off by attendance at AA meetings.

In terms of drinking history and type of drinking problem, there is abundant evidence that regular AA attenders have relatively more serious problems, usually in the mould of the traditional picture of physical addiction to alcohol, with an extremely high tolerance and the classic picture of 'loss of control' over consumption. Thus, AA is unlikely to be acceptable to those whose problems are less serious in nature or of fairly recent origin. In the past few years, AA claims to have been attracting people earlier in the course of their alcoholism, before they have reached the point described by the traditional AA concept of 'rock bottom', and also to have been responding to changes in alcoholism statistics by becoming more popular with women and young people. However, a survey of AA conducted in 1976 by David Robinson[28] showed these claims to be unfounded, at least so far as the position in Britain is concerned. It is true that, compared with an earlier survey in 1964, the number of women in AA had substantially increased and had roughly kept pace with increases in rates of female alcoholism. But what had not occurred was any reduction in the seriousness of problems shown by AA members. A more recent survey by Robinson in 1986[29] showed also that the average age of AA affiliates had remained high, with only 2 per cent under 25. All this suggests that AA continues to be a last-hope organization as far as most problem drinkers are concerned.

Does AA really work?

The question whether AA really works is one that has vexed researchers for some considerable time. Certainly, the Fellowship itself has no doubt whatever as to its own effectiveness, as the following quotation from an AA publication of 1957 shows:

> Of alcoholics who came to AA and really tried, 50% got sober at once and remained that way: 25% sobered up after some relapses, and among the remainder, those who stayed on with AA showed improvement. Other thousands came to a few AA meetings and at first decided they didn't want the program. But great numbers of these—about two out of three—began to return as time passed.[30]

In the light of research on alcoholism treatment, this rate of success is nearly miraculous, but there is the difficulty that AA self-assessments are completely unsupported by documented evidence. Despite AA claims and despite the uncritical acclamation of its effectiveness by some scientists and academics, a more objective analysis leads to different conclusions.

It is extremely difficult to undertake research with AA.[31] For one thing, the anonymity insisted on by the organization precludes the possibility of the systematic collection of research data. In addition, most attempts to evaluate AA have run into problems familiar to researchers into alcoholism treatment—interviewer bias, reliance on unsubstantiated self-reports of improvement, the confounding effects of other kinds of treatment the AA member may have gone to, the absence of suitable control groups which would make meaningful interpretation of results possible, and so on. Several studies have compared alcoholics who attend AA on a fairly regular basis with those who drop out early and have found that the former have higher rates of abstinence than the latter. However, the problems of interpretation here are obvious. It may be, for instance, that the individuals who stay longer are more motivated than others and would do better with any kind of treatment regime. Nevertheless, reviews of a number of studies suggest that abstinence rates for regular AA members vary between about 25 and 50 per cent at one year follow-up. Such rates are roughly comparable with the results of hospital treatment programmes for alcoholism.

The major difficulty in judging the true effectiveness of AA has been the absence, until recently, of research randomly allocating problem drinkers to AA and other types of treatment for the purposes of comparison. Fortunately, such a study is now available, though it gives no grounds for confidence in the superiority of AA.[32] Problem drinkers were randomly assigned to treatment by AA, individual psychotherapy by a professional, two types of behaviour therapy also conducted by professionals, and a group which received no treatment. At a follow-up one year later, it was found that, although all treatments given were better than no treatment at all, AA produced success rates inferior to those of the other treatment groups. However, two important reservations should be borne in mind here. This research dealt with alcoholic offenders referred by the courts and there are grounds for supposing that such problem drinkers are less than ideally suited to AA. The related point is that random allocation to treatment groups, although the only really satisfactory way of comparing competing treatments, is in a sense unfair to AA; as we have seen, AA may be effective with only certain types of problem drinker and ineffective with the general run of those with problems.

There is no incontrovertible evidence on which firm conclusions about the effectiveness of AA may be based, but it is possible to make some broad inferences. There is good evidence that AA is attractive to certain types of problem drinker and future research may well show that it is especially effective, compared with other interventions, with these groups. At the same time, AA has not succeeded in its aim of attracting younger problem drinkers or those whose problems are relatively less serious or of more recent origin. Despite its own claims and those of its enthusiastic supporters, there is no evidence whatever of the general superiority of AA as a treatment resource and it appears that it may even be inferior for certain types of problem drinker.

Although we have been sceptical about AA claims of success, although we have suggested that its appeal may be limited, and although we shall have occasion to be critical of its disease theory in other parts of this book, none of this should be interpreted as an attempt to denigrate the Fellowship of Alcoholics Anonymous or to detract from its achievements. For many suffering alcoholics, AA provides a source of help which is far more accessible and continuous over time than professional services are able to offer. Many thousands of alcoholics owe their lives to AA and, in the face of the general indifference and frequent hostility towards the alcoholic in our society, its efforts can only be commended.

Recent developments

In recent years AA concepts have been conscripted into commercialized treatment regimes in the United States and to an extent elsewhere, and this has become big business. Private clinics advertise on television and in the press and almost all offer lengthy in-patient stays with AA-type programmes serving as the main type of therapy. Because those attending these clinics are often funded either by private or Government insurance schemes, there is a powerful incentive among owners of the clinics to propagate the view that alcoholism is a disease; if it is not, then funders may be less likely to support long in-patient treatment programmes.

The growth of non-disease explanations of alcohol problems

in the 1960s and 1970s led to a strong counter-reaction among those with vested interests in the disease model. Strong personal attacks against the so-called 'anti-traditionalists' in the lay and professional press in the USA became common, the anti-traditionalists commonly being accused, quite fallaciously, of urging all alcoholics to drink again. There were even large law suits, only recently dropped, against some of the academics in the USA who were at the forefront of non-disease-based treatments of problem drinking.

The 'Minnesota Model' has developed over the last decade as a major form of treatment for alcohol problems. This residential group-based treatment regime is closely tied to the principles of Alcoholics Anonymous, though unlike the determinedly non-profit-making activities of AA, the Minnesota-type regimes charge large amounts to their clients. The spiritual aspects of AA are also central to this 'new' approach to the rehabilitation of alcohol problems.

The dogmatism about the nature of alcohol problems and how to treat them which haunts traditional AA approaches also figures strongly in the methods proposed by the Hazelden Foundation in Minneapolis, Minnesota, where the revised and commercialized AA approach originated. The view is put forward that only AA-type methods are effective and that recovered problem drinkers are the only truly effective agents for change. The same is said for the families of problem drinkers, as this quote by one owner of an English Minnesota-type clinic testifies:

Again, as with the primary sufferer, help by counselling is comforting but fruitless: the specific help in my view can come only from the appropriate anonymous fellowship: Al Anon or Families Anonymous.[33]

Another feature of this approach to treatment is consistent claims of abstinence rates at twelve months following treatment of between 55 and 71 per cent.[34] These figures are much higher than those obtained from all other types of treatment, but they are not substantiated by research in scientifically reputable journals.

Aside from the commercial aspects of the Minnesota-type approach, there are politico-religious threads associated with the resurgence of fundamentalism in American society. As was argued earlier in this chapter (p. 37), the roots of the AA approach are rooted in nineteenth-century revivalist religion. The growth of political fundamentalism in the United States over the last decade has coincided with a growth in all sorts of AA-type confessional movements, often framed in the language of addiction. These include 'Sexual Compulsives Anonymous' and, for their partners, 'S-Anon' ('a self-help group for those involved with sex addicts'). Shoplifters, gamblers, and 'shopaholics' are all offered AA-type twelve-step programmes, along with another estimated 200 types of 'addicts'. There is even a bi-monthly magazine on the news-stands in the USA called *Lifeline America*, started in 1988 and dealing exclusively with addictions.

Thus, in the current social climate in the USA, there can be a social cache of being a recovered 'addict', whether of alcohol or of some other drug or behaviour. The application of AA dogmas to behaviours which could scarcely be termed 'diseases'—shopping, for instance—with all the paraphernalia about recognizing these as illnesses over which one has no control, has a faintly ludicrous quality. Their acceptance by many Americans testifies to the fact that what we are witnessing here is a socio-religious phenomenon, requiring of followers the confession and repentance through which they receive status and acceptance.

Hence the attempt to explain alcohol problems (as well as other drug problems) in non-disease terms not only steps on commercial toes by threatening the theoretical basis for disease-based treatment programmes, it also threatens an entire social movement by asserting that there are other means of breaking habits than by confession and repentance. Remember the dogmatism of AA and its allied movements noted above—only AA can help the alcoholic in this view. This confirms the moral/religious/social movement basis of AA and explains why scientific data and argument are met with emotive *ad hominen* attacks: science has no say in moral/religious movements.

Many people who have overcome severe dependence on alcohol either on their own or through a non-AA-type counselling procedure are often perplexed and not a little angry at the way in which their own experience is invalidated by dogmatic assertions that only AA-type methods can lead to recovery from alcohol problems. What they fail to understand is that recovery without either professional or lay AA-type help strikes at the roots of the basis for both commercial and quasi-religious assumptions about alcohol problems.

This last point is supported by the fact that those who criticize the 'anti-traditionalists' are particularly incensed by the latter's reporting of clear scientific fact—for example, that 'minimal interventions' involving little therapist input (see Chapter 9, pp. 296–306) are on average as effective as residential treatments for moderately dependent problem drinkers, and in-patient treatment for severely dependent problem drinkers is on average no more effective than out-patient counselling. Given that many of the more vocal opponents of the anti-traditionalist lobby have personal financial interests in expensive residential clinics for problem drinking, their opposition to such ideas is understandable.

3 How many disease concepts of alcoholism are there?

In the previous chapter we traced the development of the disease view of alcoholism from its origins in the late eighteenth century to the present day. We saw that the chief motivation behind the rediscovery of the idea that alcoholism is a disease, which took place after the repeal of Prohibition in the United States, was the hope that the suffering alcoholic would thereby obtain a better deal from a largely ignorant and uncaring society. We also saw that this idea soon acquired some degree of professional and academic respectability through its endorsement by sections of the medical profession and other interested scientists. In this way, the disease perspective was put forward as a scientifically based body of knowledge, apparently grounded in the findings of medical science and supported by the full weight of medical authority. An inevitable result of this step forward, however, was that the validity of the disease perspective became a legitimate area of scientific inquiry and was forced to confront accepted rules of scientific proof and disproof. In short, the disease theory was made to enter the lions' den of scientific evaluation.

In the next chapter, we shall look at the various lines of scientific evidence that have a bearing on the proposition that alcoholism is a disease, in order to arrive at a fair and objective appraisal of the disease perspective. But before we can make this evaluation, it is necessary to prepare the ground by clarifying precisely what is meant by claiming alcoholism to be a disease.

As this suggests, there is more than one way of regarding alcoholism as a disease and this has already been implied in the historical review presented in the last chapter, when the nine-teenth-century understanding of alcohol addiction as applying potentially to anybody who drank was contrasted with the AA version which applied only to a relatively few 'alcoholics'. But apart from this distinction, there are a number of other ways in which disease theories of alcoholism have varied among themselves. In Jellinek's classic discussion of the disease theory in 1960,[1] the author was able to find no less than 115 distinct ways in which alcoholism has been described as a disease in the scientific literature up to that time.

The object here is different from Jellinek's. We are less concerned with individual disease concepts, which differ in detail, than with broad categories of theories which contain an important shared assumption or demonstrate similar general characteristics. The task is to identify these broadly different versions of the disease perspective and distinguish clearly among them, especially with respect to the manner in which the alleged cause of the condition is described.

Before we get on with this task, however, there is one other matter which it is essential to clear up. Misunderstandings frequently arise when the disease view of alcoholism is criticized and it is advisable to nip these in the bud. The problem is that it is sometimes not made clear what is *not* being denied when one denies that alcoholism is a disease. There are several connotations of the disease position which are certainly not at issue.

What is not being denied

The first fact that nobody could possibly deny is that excessive drinking causes diseases. The most common involve damage to the liver, stomach, heart, and brain but many other types of harm are possible.[2] Although excessive drinking clearly results in diseases, however, it is a completely different question whether the drinking behaviour *itself* is best regarded as a disease.

A related issue is the extent to which alcoholic drinking is a result of changes in the body's neurochemistry—in other words, to physical dependence on alcohol. Without going too far into the mechanism of physical dependence at present, we can merely say that these changes involve disturbances at the level of nervous tissue which result in withdrawal symptoms when alcohol is no longer present in the bloodstream. Dependence on alcohol in this sense is analogous to what is thought of as addiction to other drugs, such as heroin. To some extent, the drinking behaviour of the most severely dependent alcoholics *is* related to such neural changes. However, whether this is a sufficient reason for the drinking behaviour to be called a disease is again a different question.

It follows both from the physical damage caused by excessive drinking and from the partly pharmacological basis of dependence that the medical profession will always be centrally involved in society's response to problem drinking. This is the third thing we do not wish to deny. To reject the notion of alcoholism as a disease is not to reject the role medically trained personnel must assume in taking overall responsibility for the diagnosis and treatment of diseases caused by excessive drinking, and for the management of delirium tremens and other withdrawal symptoms. It is not to deny also that medical doctors have a most important part to play in the attempt to change the behaviour of problem drinkers.

Finally, it must be repeated that, in arguing against the disease perspective, we are not denying that problem drinkers deserve sympathy and compassion and may need help. Again, this is perhaps the most important message of this book—that it is possible to abandon the disease view of alcoholism without reverting to blaming and punishing the alcoholic individual.

Having hopefully cleared up some potential misunderstandings, we now turn to the promised classification of disease

Cause of the condition

	Physical	Psychological
Precedes	Type A	Type B
Follows	Type C	———

Relation to excessive drinking

Fig. 3.1. A classification of disease theories of alcoholism.

theories of alcoholism. There are three main types of disease position:

(1) alcoholism as pre-existent physical abnormality;

(2) alcoholism as mental illness or psychopathology;

(3) alcoholism as acquired addiction or dependence.

Figure 3.1 illustrates a two-way classification of these types of theory, depending on whether the cause of alcoholism is alleged to be physical or psychological and whether it is thought to precede or follow from excessive drinking. One cell of the classification in Fig. 3.1 is empty but presumably could be filled.

Type A: alcoholism as pre-existent physical abnormality

This is the most primitive version of the disease concept of alcoholism but one which continues to exert a profound influence on theory and practice in the field. Although the evidence is heavily weighted against it, it is certainly not 'a straw man', in the usual meaning of an unreal and over-simplified target of attack. The mere fact that Alcoholics Anonymous continues to propagate this kind of theory is in itself sufficient reason for taking it seriously.

The central assumption of this type of disease concept is that the alcoholic, before he or she ever comes into contact with alcohol, possesses some biochemical abnormality which leads him or her to react differently to the substance from all other human beings and which causes addiction. In the great majority of cases, the abnormality in question is seen as innate and forms part of the alcoholic's genetic endowment. Hence the importance for this type of theory of evidence relating to genetic predisposition to alcoholism (see pp. 133–40). However, an inherited abnormality is not a necessary part of these theories. The abnormality could be caused by events occurring in the uterus, resulting, for example, from excessive drinking by the pregnant mother. Whatever the original cause, all theories in this class would agree that 'one does not become an alcoholic, one is born an alcoholic'.

The AA theory of alcoholism

In the first publication by AA, the famous *Big Book* of 1939, the opening chapter is entitled 'The Doctor's Opinion'. This includes a letter from 'a well-known doctor, chief physician at a nationally prominent hospital specializing in alcohol and drug addiction', followed by an enlargement of this person's views.

The doctor in question was William D. Silkworth, whose vital role in the foundation of AA has already been described (p. 51). Dr Silkworth wrote as follows:

> We believe, and so suggested a few years ago, that the action of alcohol on these chronic alcoholics is a manifestation of an allergy; that the phenomenon of craving is limited to this class and never occurs in the average temperate drinker. These allergic types can never safely use alcohol in any form at all . . . [3]

The doctor's opinion that alcoholism was a physical and not merely a moral or psychological problem struck a chord in the experience of the founders of AA, as is shown in another quotation from the *Big Book*:

> In this statement he (the doctor) confirms what we who have suffered alcoholic torture must believe—that the body of the alcoholic is quite as abnormal as his mind. It did not satisfy us to be told that we could not control our drinking just because we were maladjusted to life, that we were in full flight from reality, or were outright mental defectives. These things were true to some extent, in fact, to a considerable extent with some of us. But we are sure that our bodies were sickened as well. In our belief, any picture of the alcoholic which leaves out this physical factor is incomplete. [3]

These quotations are about as near as AA have ever got to a formal disease theory of alcoholism. However, scattered through AA writings may be found an informal or implicit disease theory and the key characteristics of this theory all follow logically from the central idea of an allergy or some kind of pre-existent physical abnormality.

The first characteristic concerns the effects of the physical abnormality possessed by the alcoholic, whatever exactly it may be. As Silkworth says, all true alcoholics have one thing in common—'they cannot start drinking without developing the phenomenon of craving'. This craving is present from the first contact with drink, since alcohol merely potentiates the pre-existing abnormal response. Personal histories given by AA members often stress the strangeness of their very first meeting with alcoholic beverages.

While craving describes the individual's experience of an overwhelming desire for alcohol, the accompanying behaviour is known as 'loss of control', a concept which, as we have seen, was taken over from nineteenth-century temperance ideology. This does not refer merely to the fact that the alcoholic gets drunk and 'loses control' over his emotions and actions. Rather, it means that the alcoholic has become unable to choose whether to continue drinking or not. In an oft-quoted passage, Jellinek described the AA version of loss of control:

> Recovered alcoholics in Alcoholics Anonymous speak of loss of control to denote that stage in the development of their drinking history when the ingestion of one alcoholic drink sets up a chain reaction so that they are unable to adhere to their intention to 'have one or two drinks only' but continue to ingest more and more—often with quite some difficulty and disgust—contrary to their volition.[1]

Several points may be made about this short quotation. The reference to a 'chain reaction' suggests that the drinker's intentions have nothing to do with what takes place and that the only relevant consideration is the physical impact of the drug ethyl alcohol on the drinker's bodily processes. In the folklore of AA, stories are told of abstinent alcoholics being unwittingly set off on a binge by eating sherry trifle or rum pudding given by a well-meaning hostess. Moreover, the fact that 'one alcoholic drink' is all that is needed to provoke this chain-reaction leads to the well-known AA dictum 'one drink, one drunk' and this too echoes the temperance preoccupation with the evil effects of 'the first drink'. Finally, the phrase 'contrary to their volition' points to the crucial importance of the concept of loss of control in the attempt to absolve the alcoholic from moral responsibility for his actions; the alcoholic cannot be blamed for what he has done because he has lost control over his ability to choose to do otherwise. Since its revival by AA, the notion of loss of control has been a dominant theme in disease concepts of alcoholism.

A further essential characteristic of the AA model is the assertion of a hard and fast line separating the 'real' alcoholic from all other kinds of drinker. The *Big Book* devotes some space to making this qualitative distinction clear. First of all, there is the moderate drinker who 'can take it or leave it alone', something which is obviously impossible for the real alcoholic. More likely to cause confusion is 'a certain type of hard drinker' who may have the habit badly enough to lead to physical or mental impairment and may even die before his time as a result. Nevertheless, he is not to be confused with the real alcoholic because if he finds sufficiently strong reason—ill health, falling in love, change of environment, or the warning of a doctor—he can stop or moderate his drinking, 'although he may find it difficult or troublesome and may even need medical attention'. But the drinking of the real alcoholic is impervious to any sort of environmental changes and no amount of doctor's warnings can ever succeed in getting a real alcoholic to cut down permanently. Much of AA teaching is devoted to persuading the alcoholic sufferer to think of him or herself as a human being apart from others, normal perhaps in most respects when sober, but fatally different as soon as alcohol touches the lips.

We may note in passing that the AA insistence on the concept of the 'real alcoholic' is very difficult to argue against. Take two problem drinkers who, at a given point in time, appear to have a similar difficulty in controlling their intake and seem to have suffered a similar degree of damage to their lives. From the AA point of view, it is possible that one may be a 'real alcoholic' and the other merely a 'hard drinker'; there is no way of telling which until we discover what is going to happen to them. If one does succeed eventually in controlling his drinking, perhaps by finding a happy marriage or changing to a job which does not involve drinking with clients, he is

clearly the hard drinker; if the other experiences similar kinds
of life-events but nevertheless goes on to destroy himself, he is
obviously the real alcoholic. But there is no way of knowing
this until after the event. For this reason, the concept of the
real alcoholic cannot help us to predict what is going to happen
in the future and, by the same token, becomes impossible to
disprove, since anyone who appears to have the attributes of
the real alcoholic but who does in fact control his drinking can
be said to be, by definition, not a real alcoholic to begin with.
Concepts like this are called 'tautologous' and are useless for
scientific purposes because they cannot add to our understanding of the world.

It also follows from the allergy idea that the disease of
alcoholism is irreversible. It cannot be cured but only be
arrested by total and lifelong abstinence; in the words of the
best-known AA maxim of all, 'once an alcoholic, always an
alcoholic'. AA members continue to refer to themselves as
alcoholics even after a lifetime's successful avoidance of alcohol
and the outsider at an AA open meeting can never fail to be
surprised when the long-standing member begins his tale with
the words: 'My name is John and I'm an alcoholic.' In AA
terminology, one is never a recover*ed* but only a recover*ing*
alcoholic.

In AA writing the hope is sometimes piously expressed that
scientists will one day discover the exact nature of the mysterious allergy and thereby come up with a permanent cure for the
disease. But this is not presented as a realistic possibility and
one has the impression that no committed AA affiliate seriously
expects it to occur in his lifetime. Indeed, much of the initial
response to the new recruit is concentrated on getting him to
accept fully the uncomfortable truth that he will never be able
to drink again and to helping him settle down to an existence
in which alcohol plays no part. He is told that alcoholism is not
only an irreversible disease, it is also progressive, so that any
continued drinking can only lead to further deterioration and

death. This, then, is the final characteristic of the AA model which follows on from the notion of an incurable allergy to alcohol and which has also been inherited from nineteenth-century thought—the absolute insistence on total and lifelong abstention from alcohol as the only path to recovery.

Early nutritional and endocrinological theories

Although supported by a medical doctor, the AA notion of an allergy to alcohol was clearly an amateurish guess at the nature of the crucial predisposing factor. It was to be expected, then, that when things were getting back to normal after the end of the Second World War scientists should come up with some more likely-looking candidates.

Perhaps the best-known of these early disease concepts is the so-called 'genetotrophic' theory of R. J. Williams which was popularly described in *Scientific American* in 1948[4] and attracted great publicity at the time.

His theory begins with the observation that there is considerable variation in human metabolism, particularly in the activity of the enzymes which control chemical activity in the body. It then goes on to say that these differences in individual metabolism are genetically transmitted and that many disorders of metabolism are hereditary. Thus, genetic factors are thought to lead to the under-production of one or more specific enzymes which leads in turn to the inability of the body to utilize the corresponding nutritional elements, notably vitamins. For this reason, the individual possesses an extra 'cellular need' for certain vitamins and, given that these vitamins are obtainable in alcohol, an insatiable craving for that substance will develop and predispose the individual to alcoholism.

The next chapter will survey evidence for or against theories of physical abnormality in general—for instance, evidence on the inheritance of alcoholism and on the relative importance of cultural as opposed to biological factors in determining rates of problem drinking. Here it will be instructive briefly to follow the particular fate of the genetotrophic theory, although this is now of historical interest only. It serves as a classic example of a great many other theories of physical abnormality which have been put forward with much excitement as finally solving the problem of alcoholism, only to fall by the wayside as a consequence of empirical research.

Unfortunately for the genetotrophic theory, the only evidence which could be used to directly test it was based on experiments with animals, usually mice or rats, since it is obviously impossible for ethical reasons to manipulate experimentally the vitamin balance of human beings. In the first experiments taken to support the theory, it was found that rats which suffered a deficiency of vitamin B increased their intake of alcohol when given a choice between alcohol and water. Also, certain strains of rats were more likely to respond to vitamin deprivation in this way than others, thus implicating genetic factors.

But, of course, this is a far cry from a demonstration of the postulated genetotrophic mechanism in human alcoholics. While Williams speculated about the possibility of being able to pick out potential alcoholics by means of biochemical tests, no consistent differences emerged when the metabolic processes of diagnosed alcoholics and normal subjects were compared. What is more, a craving or need for alcohol has never been shown to exist in the pre-alcoholic state of alcoholics. Finally, even the animal experimentation turned against the theory when it was shown that, if a sweet solution were added to the choices offered to vitamin-deprived animals, their increased intake of alcohol disappeared.

Although this theory has been discredited and no longer stimulates research, it should not be assumed that the search for a metabolic basis to alcoholism has been abandoned. On the contrary, scientific papers in this tradition are still frequently found in the international network of learned journals devoted to alcoholism research.

More recent genetic theories

Of the many subsequent theories that have been proposed to account for the alleged genetic predisposition to alcoholism, two which attracted attention during the 1960s merit separate consideration. The first stems from the bizarre and seemingly irrelevant observation by Chilean researchers that alcoholics show a greater incidence of a certain type of colour-blindness than normal control subjects.[5] The significance of this is that it suggests that colour-blindess is a 'genetic marker' for the disease of alcoholism, both phenomena being consequences of the same underlying physiological malfunction. This was hypothesized to involve an abnormality of liver metabolism, transmitted via a sex-linked recessive gene, thus accounting for the much lower incidence of alcoholism in women than men. On a less precise level, the association between colour-blindness and alcoholism is encouraging to those who believe the latter to be inherited, since it shows it to be correlated with a condition which undoubtedly is genetically transmitted.

Apart from more general evidence on the inheritance of alcoholism which will be dealt with in the next chapter, the colour-blindness hypothesis has been criticized on two grounds. Firstly, the authors' understanding of colour-vision physiology has been shown to be faulty and, secondly, attempts to replicate the finding of the key association between alcoholism and colour-blindness in Britain and the United States have simply failed to confirm it.

The second popular line of research during the 1960s concentrated on possible differences in the blood groups of alcoholics and normal subjects. The original finding from a hospital in Colorado claimed that alcoholics were more likely to be of Blood Group A. As with so many chance findings of this sort, however, subsequent research has failed to confirm the original observation. A further problem with the blood-group research is that it is difficult to eliminate the possible confounding effects of differing racial origins between groups of alcoholics and controls.

Type B: alcoholism as mental illness or psychopathology

Theories of this type are similar to those in Type A, in that they assume some individuals to be predisposed to develop alcoholism (see Fig. 3.1). However, the predisposing factor here is not a physical, but rather a psychological abnormality and explanations are founded on such central concepts as 'mental illness', 'emotional illness', 'psychopathology', 'neurosis' and 'personality disorder'. Where in Type A theories the preoccupation of the researcher is the hunt for the physical abnormality differentiating the alcoholic from all other drinkers, here the search is for the crucial 'alcoholic personality' which results in a special vulnerability to dependence.

Psychodynamic theories

Perhaps the most important contribution to this class of theory has come from psychoanalysis and related psychodynamic formulations. Unfortunately, there are almost as many psychodynamic theories of alcoholism as there are psychoanalysts who have addressed themselves to the topic. Nevertheless, it is possible to pick out certain themes in the literature which have given rise to broad traditions of research and therapy. The common ground for psychodynamic explanations is the assumption that alcoholism results from unconscious impulses which have been repressed but which find symbolic expression in heavy drinking.

The founder of psychoanalysis, Freud himself, was responsible for two separate accounts of the origins of alcoholism. In *Mourning and melancholia* published in 1917,[6] he suggested that, as a consequence of defects in the relationship between parents and child, the alcoholic was fixated at the earliest stage of psychosexual development, the oral stage. This suggestion gave rise to a prominent research tradition in which the alcoholic is seen as having an 'oral personality', characterized primarily by a lack of self-control, passive dependence on others, self-destructive impulses, and a tendency to use the mouth as a primary means of gratification. Drinking provides such a means of gratification and intoxication allows the expression of other features of the oral personality. As with all accounts of psychopathology based on psychosexual fixation and regression, the alcoholic was regarded as inadequate to meet the demands of everyday reality and as using alcohol to regress to a childhood level of thinking in which gratification in fantasy was unhampered by the constraints of logic. The simplified notion that the alcoholic uses alcohol as an 'escape from reality' has become commonplace among treaters and sufferers from drinking problems.

The hypothesis of the oral personality has proved extraordinarily difficult to test. Not only do authorities differ in their account of the basic personality type but some writers describe the future alcoholic's mother as over-indulgent and over-protective while others assert that she is cold and neglecting. Apart from this kind of inconsistency, research in this area is of generally poor quality, many studies not even using a control group of normal, non-alcoholic subjects to provide a comparison with the personalities of alcoholics. Moreover, it is not clear why the psychopathology described is thought to be specific to alcoholism and not also involved in other forms of deviant behaviour. Similar kinds of explanations have been proposed for smoking and, more generally, for so-called character disorders and sociopathic personalities. Finally, it is possible that the regression to a more primitive level of thinking which is held to be a feature of the alcoholic's drinking experience is also true of many others who may occasionally become intoxicated but do not suffer from obvious drinking problems.

Later, in 1930, Freud revised his theory of alcoholism.[7] On this occasion he used the idea of repressed homosexual impulses which he had previously employed to explain paranoid symptoms. Freud maintained that the alcoholic found expression for unconscious homosexual wishes in the drinking company of other men. Disappointment in relationships with women drove the alcoholic to drink, but whenever the latent homosexuality rose to the surface, he would return to his wife and immediately suspect her of sexual activity with those men he unconsciously loved. Here Freud must have been impressed by the morbid jealousy which is occasionally found in the behaviour of alcoholics but, although this may apply to a few individual cases, it is difficult to imagine it as a general explanation of alcoholism.

One other psychodynamic theory which became very popular in its day and still has its adherents should briefly be mentioned. In his best-selling book *Man against himself* published in 1938, Karl Menninger proposed that the alcoholic has a powerful but unconscious desire to destroy himself.[8] This desire originates in the child's feeling of being betrayed by his or her parents and the frustration arising from this results in an intense rage towards the parents which comes into conflict with the child's fear of losing them. In later life, repressed hostility towards the parents creates a need for self-punishment to alleviate feelings of guilt. Hence the slow suicide of alcoholism. As with other psychodynamic formulations, Menninger's theory has touched the popular imagination and finds echoes in autobiographic accounts of alcohol dependence but, again, it is difficult to understand how it could lead specifically to excessive drinking and not to other forms of harmful behaviour.

The personality-trait approach

The search for the elusive 'alcoholic personality' has also been conducted from a quite separate starting-point from psychoanalysis. This is the so-called trait psychology in which theories of personality organization are based on the results of huge numbers of questionnaires designed to measure differences in the way people behave, or at least *say* they behave. The basic assumption is that there exist universal dimensions of personality, present to a varying extent in all people, which have a fixed and constant effect on their behaviour in all manner of situations. Using questionnaires derived from trait theories, vast numbers of scientific papers have been published reporting attempts to find consistent differences between the personality structures of alcoholics and non-alcoholics, differences which are then assumed to have causal significance in the development of alcoholism.

It will be impossible to summarize all this evidence here. What can be said is that this kind of research is bedevilled by the familiar difficulty of distinguishing the causes from the effects of alcoholism when differentiating features of alcoholics are isolated. The only really satisfactory solution to this problem is to conduct longitudinal studies in which personality characteristics measured in general populations, usually children, are subsequently examined to see whether they can predict alcoholism at a later date. There is some evidence from such longitudinal studies that people diagnosed as alcoholics in adulthood may have shown a tendency to get into trouble with authority in childhood.[9] However, this hardly amounts to a specific type of personality and, of course, there are many other diagnosed alcoholics and problem drinkers who did not get into trouble as children.

Inferiority and powerlessness

In addition to Freudian theories of alcoholism, the Adlerian school of individual psychology, which is centred on the idea of the 'inferiority complex', has also had its say on the subject. To

Adler and his followers, feelings of inferiority, caused by over-indulgent mothering and a consequent inability to face up to the frustrations of adult reality, were at the root of all drinking problems. Alcohol enabled the individual to abolish feelings of inferiority while at the same time evading responsibilities.

A similar but more modern approach was adopted by McClelland and his associates.[10] In this view, the alcoholic is someone who has a greater than normal need for power but inadequate resources of personality to achieve his ambitions. In the face of this frustration, the individual uses the effects of alcohol to achieve a temporarily euphoric sense of power and achievement while simultaneously discharging accumulated anxiety and tension. Because drinking reduces the alcoholic's problem-solving abilities still further, this creates a vicious circle in which his problems mount and his feelings of power-lessness become even greater, with an accompanying increase in the need for alcohol. McClelland's thesis is consistent with a general view of alcoholism which implicates the special problems of the male in a society which sets great store by masculine success and power.

In support of the striving-for-power hypothesis, there is evidence that some alcoholics say they drink because it helps them to feel superior. However, the experiments reported by the McClelland team were all conducted with normal subjects. They were shown ambiguous drawings after ingestion of alcohol and an increase in power themes in their interpretations of the drawings was typically recorded. It is true that a proportion-ately greater increase in power fantasies occurred among the more heavy-drinking non-alcoholic subjects but there is as yet no evidence that a similar increase would be found in the fantasies of the majority of alcoholics.

Type C: alcoholism as acquired addiction or dependence

We come now to the most modern and, at least in professional circles, currently most influential type of disease concept. In contrast to both previous types, this point of view assumes that the disease the alcoholic suffers from does not precede his harmful drinking but is a consequence of it (see Fig. 3.1). Thus, there is no special predisposition to contract the disease and anyone who tries hard enough can become an alcoholic. By a twist of historical fate, the idea of the founders of the Temperance Movement, that it was the addictive nature of alcohol itself which was responsible for the disease of alcoholism, has once more achieved prominence.

Jellinek of the phases

E. M. Jellinek's work continues to exert a major influence on the alcoholism field. His first important contribution was a journal article published in 1952 entitled *Phases of alcohol addiction*.[11] This seminal piece of research was heavily influenced by AA views and it is sometimes suggested that it was merely an attempt to dress up the AA model in academically respectable clothes. This contains some truth but is not the full story.

It is true that Jellinek's questionnaire on the course of drinking problems was filled in entirely by AA members after having been distributed through the official AA publication, *Grapevine*. Even the contents of the questionnaire were constructed by AA members. It is not surprising in view of this that the phases of alcoholism which emerged—pre-symptomatic, prodromal, crucial, and chronic—and the progression of events which made up the alcoholic's slide to 'rock bottom', were a reflection of AA folklore. Later on, the symptoms described in Jellinek's progression of phases were cast by Max

Glatt into the U-shaped chart shown in Fig. 3.2.[12] This 'slippery slope' into alcoholism and the upward path to recovery has become one of the most familiar artefacts of the Alcoholism Movement. It can be argued, however, that its form has more to do with eighteenth- and nineteenth-century moral warnings against the dangers of alcohol, such as Hogarth's 'Rake's Progress', than to any body of scientific support. Glatt's chart may be compared with Benjamin Rush's 'Moral and Physical Thermometer' in Chapter 2 (p. 31).

The fundamental distinction drawn by Jellinek between alcohol addicts properly so-called and 'habitual symptomatic excessive drinkers' clearly echoed the AA differentiation between 'real alcoholics' and mere 'hard drinkers'. It was only to alcohol addicts that the disease concept applied and the main defining characteristic was loss of control, which occurred only among alcohol addicts and only after many years of excessive drinking. Jellinek did not deny that the habitual symptomatic excessive drinker was also a sick person, but 'his ailment is not the excessive drinking but rather the psychological or social difficulties from which alcohol gives temporary surcease'. These difficulties were originally present in both kinds of problem but with alcohol addicts the process of loss of control was superimposed on psychological pathology. However, whether this superimposed loss of control is itself psychological or physical in nature was not known. Neither has it been established, says Jellinek, whether loss of control originates in 'a predisposing factor X'—an obvious reference to the AA allergy factor—or is acquired in the course of excessive drinking. Here is the break with AA dogma. Although his research confirmed the AA portrayal of the progressive course of the disease, Jellinek is studiously non-committal about the origins of loss of control.

Jellinek of the species

Jellinek's second major contribution, a book entitled *The disease concept of alcoholism*, was heavily influenced by the international experience he had acquired through his association with the World Health Organization. He was impressed by the many different types of alcohol-related behaviour which were identified as problems in different parts of the world and concluded: 'By adhering strictly to our American ideas about "alcoholism" and "alcoholics" (created by AA in their own image) and restricting the term to those ideas, we have been continuing to overlook many other problems of alcohol which need urgent attention'.[1] For this reason he proposed a wide-ranging definition of alcoholism: 'any use of alcoholic beverages that causes any damage to the individual or society or both'. But because this definition was broad and vague, Jellinek went on to make a clear distinction between 'alcoholism' and 'alcoholics'; alcoholics were not the same as those who suffer from alcoholism in the broad sense just described, but were confined to those fitting the species of alcoholism which were properly called diseases.

Of the five species of alcoholism identified by Jellinek and labelled with Greek letters, only two, the gamma and the delta species, were definitely described as diseases. Gamma alcoholism is the term used to refer to what Jellinek in his earlier work had called simply 'alcohol addiction' and is the species which was moulded in the image of AA. It has four defining characteristics: acquired increased tissue tolerance; adaptive cell metabolism; withdrawal and craving; and loss of control. It is the predominating species in Anglo-Saxon countries and the one causing the highest degree of damage in health, financial, social, and interpersonal areas. Delta alcoholism, on the other hand, is the species associated with the inveterate drinking found in France and other wine-drinking countries. It shows the first three characteristics of gamma alcoholism just listed, but instead of 'loss of control' there is 'inability to abstain'. The ability to control intake on any given occasion remains unaffected but there is no capacity for abstention even for a few days without the occurrence of withdrawal symptoms.

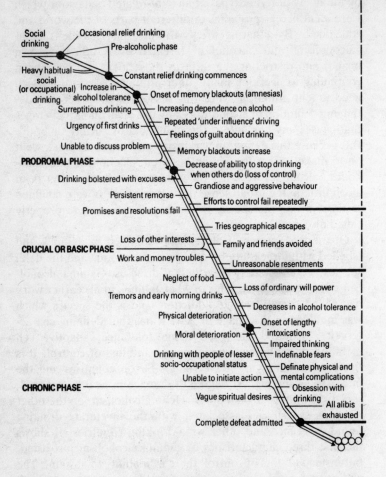

To be read left to right

Social drinking

Occasional relief drinking

Pre-alcoholic phase

Heavy habitual social (or occupational) drinking

Increase in alcohol tolerance

Constant relief drinking commences

Onset of memory blackouts (amnesias)

Surreptitious drinking

Increasing dependence on alcohol

Urgency of first drinks

Repeated 'under influence' driving

Unable to discuss problem

Feelings of guilt about drinking

Memory blackouts increase

PRODROMAL PHASE

Decrease of ability to stop drinking when others do (loss of control)

Drinking bolstered with excuses

Grandiose and aggressive behaviour

Persistent remorse

Efforts to control fail repeatedly

Promises and resolutions fail

Tries geographical escapes

Loss of other interests

Family and friends avoided

CRUCIAL OR BASIC PHASE

Work and money troubles

Unreasonable resentments

Neglect of food

Loss of ordinary will power

Tremors and early morning drinks

Decreases in alcohol tolerance

Physical deterioration

Onset of lengthy intoxications

Moral deterioration

Impaired thinking

Indefinable fears

Drinking with people of lesser socio-occupational status

Definate physical and mental complications

Unable to initiate action

Obsession with drinking

CHRONIC PHASE

Vague spiritual desires

All alibis exhausted

Complete defeat admitted

Fig. 3.2. Glatt's chart of alcohol dependence and recovery [reproduced with permission from Glatt, M. M. (1982). *Alcoholism*. Hodder and Stoughton, London].

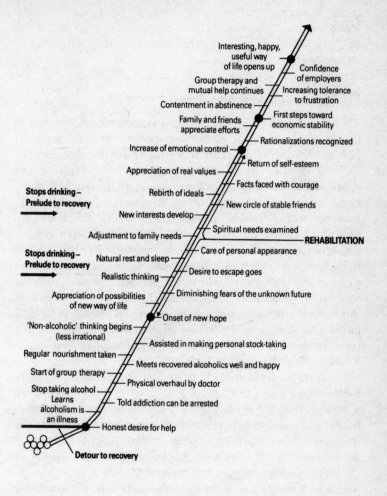

Interesting, happy, useful way of life opens up

Confidence of employers

Group therapy and mutual help continues

Increasing tolerance to frustration

Contentment in abstinence

First steps toward economic stability

Family and friends appreciate efforts

Increase of emotional control

Rationalizations recognized

Appreciation of real values

Return of self-esteem

Rebirth of ideals

Facts faced with courage

New interests develop

New circle of stable friends

Stops drinking – Prelude to recovery

Adjustment to family needs

Spiritual needs examined

REHABILITATION

Natural rest and sleep

Care of personal appearance

Stops drinking – Prelude to recovery

Realistic thinking

Desire to escape goes

Appreciation of possibilities of new way of life

Diminishing fears of the unknown future

Onset of new hope

'Non-alcoholic' thinking begins (less irrational)

Assisted in making personal stock-taking

Regular nourishment taken

Meets recovered alcoholics well and happy

Start of group therapy

Physical overhaul by doctor

Stop taking alcohol

Told addiction can be arrested

Learns alcoholism is an illness

Honest desire for help

Detour to recovery

According to Jellinek, the reason that only the gamma and delta species can be considered as diseases is that only they entail the 'physiopathological changes' analogous to those found in drug addiction. These changes are responsible for bringing about craving and either loss of control or inability to abstain, as the case may be.

The alcohol-dependence syndrome

There is no doubt which is the most significant recent event in the development of the disease perspective. It is the formulation in 1976 and in the years following of the 'alcohol-dependence syndrome' by Griffith Edwards, now Professor of Addiction at the Institute of Psychiatry, and the late Milton M. Gross. As a consequence of an impressive body of work, Edwards would now be recognized by many as the world's leading authority on alcohol problems and his transatlantic collaboration with Gross, who himself possessed a distinguished track-record as a biomedical researcher in the field, was bound to be highly influential. Moreover, despite the fact that the original, 1976 delineation of the syndrome[13] was described as 'provisional', it was approved by the World Health Assembly in the very same year it was first mooted and became officially incorporated into the International Classification of Diseases as a new medical diagnosis in January 1979.

We shall not have space here to describe the seven separate syndrome elements but a section of Chapter 5 is given over to a demonstration that each of these elements can be meaningfully translated into behavioural concepts which are not specific to alcohol dependence and which, indeed, need not assume a drug-involvement of any kind. In doing so, a fuller account of the set of phenomena covered by the syndrome is provided.

Edwards and Gross begin their description of the syndrome by conceding that it is based more on clinical impression than substantiated scientific evidence. The authors were also quick to make clear what they meant by the term 'syndrome'. It refers merely to 'an observable coincidence of phenomena'. For this coincidence to qualify as a syndrome, not all the described phenomena need always be present, nor present to the same degree. Thus, in Edwards's and Gross's conception, the alcohol-dependence syndrome is subtle and changing in appearance.

The further question of whether or not the syndrome is to be regarded as all-or-none is a vexed one. In earlier writings, it seems that it applied, with varying intensity, to the entire drinking population. At other times, however, it is presented as applying only to a special segment of serious problem drinkers. In later work, Edwards made it perfectly clear that the dependence syndrome is 'a very abnormal condition indeed'.[14] Although some kind of dependence can be said to be present, in a weak kind of sense, among nearly all drinkers, the syndrome is quite separate from normal behaviour and experience.

Another feature of the syndrome is that it abolished the distinction made by Jellinek between physical and psychological dependence. Instead, the syndrome is to be viewed as a 'psycho-physiological disorder' in which the contributions of psychological and physical factors cannot meaningfully be separated. In line with the extreme plasticity of the syndrome, Edwards stresses that it is always 'environmentally and personally coloured'. Its appearance will be markedly different in the heavy-drinking milieu of France from the dry American state of Utah. Equally, it will show itself in a very different form in a person of poor impulse control than in a previously well-integrated personality and many other factors associated with personality or individual social circumstances will affect the form it takes. Nevertheless, underlying all these personal, social, and cultural variations, there is the reality of a single core syndrome.

In the years following its original publication, the syndrome idea has provoked a lively and sometimes heated debate. One of its foremost opponents is the sociologist Stan Shaw and some of his criticisms will briefly be mentioned.[15] First, Shaw charges that the presentation of the syndrome is full of ambivalence, not only with respect to the issue of whether or not it is all-or-none, but also with regard to the exact nature of the dependence said to be involved. Second, he claims that the described syndrome is full of tautology and rests on fundamentally circular arguments. Third, he asserts that the only evidence which has been produced to support the syndrome concept is that pertaining to 'altered psycho-biological state' and to increased tolerance, withdrawal and the reinstatement of these after abstinence. No one could seriously dispute the existence of a syndrome of withdrawal—or, in more recent terminology, a state of 'neuro-adaptation' to alcohol. It is only when aspects of feeling and behaviour are tacked on to the withdrawal syndrome to form the larger idea of a core dependence syndrome that the evidence fails to be in its favour.

Another of Shaw's criticisms is immediately relevant to the task of classifying disease concepts of alcoholism. This is the point that the leading symptom of the syndrome is 'impaired control over the drug ethyl alcohol'.[16] Given that the syndrome is admittedly all-or-none, given that the recommended way of measuring it is confined to tolerance, withdrawal symptoms, relief drinking, and reinstatement, and given that it is the presence of physical dependence which seems essential to make the diagnosis, this represents little advance over Jellinek's second disease concept described in 1960 (see pp. 80–4).

There is some equivocation about whether the syndrome is to be regarded as a disease at all. For instance, in 1977, Edwards wrote that 'whether a syndrome is a disease is largely semantic'. What this ignores is that labelling a problem as a disease has profound consequences for how society organizes its response to that problem and, equally as important, for how those who have the problem make sense of their own behaviour and go about seeking solutions to it. There can be little doubt that the word 'syndrome' carries heavy disease-like overtones and suggests that specifically medical treatment will be needed. In later writings, Edwards appears to have acknowledged the continuity between the syndrome and earlier disease concepts when he wrote that 'it is an idea roughly coterminous with what many people would call the disease of alcoholism, or with the Alcoholics Anonymous notion of what counts as alcoholism'.

In view of the haste with which a provisional syndrome was rushed on to the world-wide alcoholism stage and in view of the absence of any convincing evidence to support it, Stan Shaw has argued that the development of the alcohol-dependence syndrome was a political rather than a scientific achievement:

. . . the syndrome idea was not simply an attempt to find a substitute for the concept of alcoholism; rather it was an attempt to create a particular kind of substitute concept—one which coped with all the critiques of the disease theory of alcoholism, yet which retained all its major assumptions and implications. The dependence syndrome concept would be a rationale for those persons involved with dealing with drinkers, who were perhaps embarrassed by the scientific untenability of the strict disease theory, but who still felt nevertheless that people with drinking problems were ill, that their essential trait was an inability to control alcohol intake and therefore that the basic treatment goal of abstinence should be retained in most cases.[15]

A syndrome of disease theory assumptions

To conclude this chapter, we shall isolate a set of assumptions which are common, to a greater or lesser extent, to all disease theories and which characterize the disease perspective on problem drinking as a whole. This is necessary to prepare for the evaluation of the relevant scientific evidence presented in the next chapter.

At the risk of seeming flippant, an analogy with the form taken by the alcohol-dependence syndrome, as described by Edwards and Gross, will be useful for this purpose. As with the elements of the syndrome, we do not claim that all the following assumptions are present in every single disease concept, nor do we suggest that they are all described in the same way or given the same weight. However, they occur together sufficiently often to form a general characterization of the essential features of the disease perspective. The assumptions are four in number:

1. *Alcoholism is a discrete entity*. The basic assumption is that the drinking behaviour and alcohol-related experiences of people suffering from the disease of alcoholism are qualitatively distinct from the behaviour and experiences of those not suffering from the disease. All earlier disease theories subscribe to this assumption and, once the precise nature of the alcohol-dependence syndrome became clear, it too emerged as a discrete entity.

The corollary of this assumption is that the disease of alcoholism is a *single* entity. It is true that Jellinek defined at least two, independent types of disease but, despite the prestige of Jellinek's work, this has not gained wide acceptance. In the alcohol-dependence syndrome, the two types have been collapsed back into a single core construct of underlying pathology.

2. *Alcoholic drinking results from an involuntary impaired control over drinking and an abnormal craving for alcohol*. The outstanding reason given in earlier disease theories for the fact that the alcoholic's drinking behaviour is qualitatively different from normal (assumption 1) is that he or she experiences an uncontrollable craving or compulsion to drink alcohol, and has lost control over the amount and/or frequency of drinking. In later, more sophisticated positions, it is admitted that craving does not always result in uncontrolled drinking and that control

need not be completely lost but merely 'impaired'. Nevertheless, some form of craving and diminished control over drinking is invariably presented as the nature or essence of the condition.

The effect of this second assumption, in common with all similar assumptions used in psychiatry, is that the alcoholic's behaviour is wholly or partly involuntary and not completely within the realm of personal choice. This is contrasted with the drinking behaviour of non-alcoholics which is under voluntary control and fully within the realm of choice.

3. *Impaired control and craving are irreversible.* The impaired control and craving which are the cardinal symptoms of alcoholism (assumption 2) are assumed to be irreversible. Alcoholism cannot be cured, only arrested. This is due to pre-existent biochemical or psychological abnormalities in the alcoholic which make sustained normal drinking impossible or to presumably permanent neurophysiological changes brought about by years of excessive drinking. This irreversibility results in the outstanding implication of the disease perspective for treatment—that the diagnosed alcoholic must maintain a total and lifelong abstinence from alcohol.

4. *The irreversibility of the disease of alcoholism (assumption 3) entails a progressive deterioration in the alcoholic's condition if drinking is continued.* Earlier disease concepts described a relatively inexorable sequence of symptoms or phases of alcoholism. Later positions may not subscribe to an invariable sequence but nevertheless assume that alcoholism, like any other disease, shows an inherent natural history and course which, if left untreated, leads to further deterioration. Treatment is therefore essential to the alcoholic's recovery.

4   The evidence

We have seen that the twentieth-century disease theory of alcoholism began to look suspect when it emerged from the relative obscurity of its origins in the Alcoholism Movement and was subjected to the light of scientific scrutiny. Here we will examine the various lines of evidence which are relevant to an evaluation of the disease concepts of alcoholism described in the last chapter. We make no bones, however, about our ultimate intention in this chapter, which is to show that the disease theory simply cannot handle the facts.

We begin with a line of inquiry which has had perhaps the most dramatic impact on the alcoholism field in recent times and which has certainly led to the greatest amount of controversy. This evidence is a fine example of the Kuhnian anomaly—the scientific finding which cannot be explained by an existing paradigm, and which has the effect of subverting it and leading eventually to its downfall in a scientific revolution (see pp. 3–6). The evidence in question is the repeated finding that persons diagnosed as 'alcoholics' or 'alcohol addicts' can sometimes return to normal, controlled patterns of drinking.

Normal drinking by former alcoholics

On a day in 1958 at the Maudsley Hospital in London, the leading teaching and research centre of British psychiatry, a certain Dr D. L. Davies was in conversation with the social worker attached to his treatment team, Edgar Myers. Davies was in charge of four or five beds for alcoholics but was by no means exclusively interested in alcoholism, preferring to see it as an integral part of general psychiatry. Myers' job was to organize and carry out the follow-up of all alcoholics treated in the unit.

During this conversation, Myers remarked that there was a group of people who were drinking but appeared not to show any social difficulties. Being aware of the prevailing disease-theory logic, Myers said, 'You realize these people shouldn't happen?' Davies replied, 'Well, they have happened!'

In this informal and accidental way began a piece of research which was to have profound consequences for the modern understanding of alcohol problems. Possibly because of his lack of specialization in the field, which may have given him a degree of immunity from the dogma of abstinence as the only possible solution to a drinking problem, Davies decided to take a closer look at these former alcoholics who were allegedly drinking in a normal fashion. They were all visited by Davies himself at home and sometimes at work, and specific inquiries were made, of relatives as well as the ex-patients themselves, about drinking since discharge from treatment. To qualify for inclusion as a normal drinker, the patient had to show at least five years' harm-free drinking and to have been diagnosed as an 'alcohol addict', with the cardinal symptom of 'loss of control' over drinking. Using these criteria, Davies was able to find just seven men out of ninety-three patients followed up.

Davies's first attempt to have a paper published reporting his follow-up was rejected by the *Lancet* on the grounds of insufficient general interest and it eventually appeared in the *Quarterly Journal of Studies on Alcohol* in 1962.[1] Because of his comparative innocence, at the time, of the depths of emotionalism in the alcoholism field, Davies can scarcely have anticipated the extraordinary reaction to his brief and cautiously worded report. The journal was deluged with letters of protest and subsequently published an unprecedented eighteen commentaries from experts throughout the world. The great majority were highly critical of Davies and his research, although he did receive some support.

Since the controversy found its way for the first time into the popular press, several commentators expressed concern lest abstinent former alcoholics should come to hear of Davies's findings and take it upon themselves to experiment with alcohol too, with possibly disastrous consequences. In his reply, Davies was most sympathetic to this anxiety but reminded his critics that he had not recommended any change of treatment goal on the basis of his findings and had emphasized that all alcoholics should still be aimed at total abstinence. He also pointed out that, although communications in professional journals might appear in the press with unwelcome repercussions, this could never be an argument for suppressing scientific findings. This point was reinforced by one contributor to the debate who recounted that, when he had reported in 1957 that it was possible for confirmed alcoholics to return to social drinking,[2] he had been virtually ordered by the body funding his research to omit these 'embarrassing' findings.

As Davies had been careful to point out, he was not the first to report the occurrence of resumed normal drinking. Indeed, it is not at all clear why Davies's paper caused such a fuss when previous examples had not. It may have been because he was the first to draw attention to the phenomenon in the title of his article. It is also possible that it appeared at just that time when the disease theory was beginning to attract widespread support and aroused hostility because it brought under fire the key assumption of irreversibility.

The most important conclusion to be drawn from the early reports is that normal drinking outcomes following treatment for abstinence began to be observed as soon as even moderately well-designed follow-up studies were conducted. This is import-ant to note because proponents of the disease perspective sometimes give the impression that normal drinking in former alcoholics is something which has only just been dreamt up by naïve and inexperienced researchers. Nothing could be farther from the truth and this particular inadequacy of the disease theory became obvious from the first introduction of scientific methods in the field.

After the immediate controversy surrounding Davies's 1962 paper had subsided somewhat, it was natural that other researchers should turn their attention to this crucial topic. In the remaining years of the decade, a handful of articles appeared which thoroughly explored and corroborated the main implications of Davies's original findings. One important study by Kendell,[3] also carried out at the Maudsley Hospital, resolved some of the ambiguities in Davies's paper by demonstrating that there was no upper limit to the level of dependence which would absolutely preclude a normal drinking outcome.

The conclusions which may be drawn from all this research will be left to the summary at the end of this section (pp. 107–8). Here one general point may be made before passing on. A count of studies reporting normal drinking in former alcoholics conducted in 1979[4] found over eighty. This count was unavoidably incomplete and, in any event, the number has undoubtedly increased by now. Nevertheless, it is still possible to find people involved with alcoholism treatment who deny that resumed normal drinking is possible. This does not apply only to members of AA but includes eminent psychiatrists and other professionals working in the field. These individuals simply declare that it is all dangerous nonsense and that no such persons exist.

The problem here is to explain how it is that sincere and devoted therapists have failed to observe resumed normal drinking when the research evidence has been so overwhelmingly in its favour. One reason often proposed for this is that patients who successfully maintain normal drinking are unlikely to keep in contact with treatment facilities where the goal of total abstinence has been strongly urged on them and the possibility of recovered control ridiculed. It should always be remembered that resumed normal drinkers are often acting in defiance of the medical advice they have received. There is indeed some evidence to support the idea that abstinence-oriented treatment personnel are unlikely to come into contact with resumed normal drinkers. Beyond this, it is also possible that normal drinking outcomes are not registered, precisely because the category does not exist in the observer's conceptual framework. It is difficult to overestimate the subtle effects of prior assumptions and ideological commitments. But in the last resort, there is no conceivable way these sweeping assertions that resumed normal drinkers have never been encountered could ever be falsified; there is no way, in other words, they could ever be proved or disproved by collecting evidence. Therefore, no matter how experienced or distinguished the authority uttering these assertions, they can have no place in serious scientific inquiry.

The Rand Report

Despite the fact that it had been replicated many times during the years since D. L. Davies published his article in 1962, the finding of normal drinking in former alcoholics again became the subject of a furious controversy in 1976, following the appearance of the so-called Rand Report.[5] If anything, this controversy was even more bitter and nation-wide media in America were more fully involved.

In 1971 the United States government founded the National Institute on Alcohol Abuse and Alcoholism (NIAAA). As part of its remit, NIAAA established a network of Alcoholism Treatment Centers (ATCs) throughout the country and provided for the continuous monitoring of their effectiveness. The Rand Corporation, a private organization specializing in social-science research, was commissioned to collate and analyse all the data collected between 1970 and 1974 relevant to the outcome of treatment given by the ATCs. The Rand Report was the first statement of findings from this vast research effort, representing by far the largest follow-up of treated problem drinkers ever conducted, with 2339 male alcoholics interviewed six months after treatment and 589 eighteen months later.

In the summary section of the report, the authors stated that their findings suggested 'the possibility that for some alcoholics moderate drinking is not necessarily a prelude to a full relapse and that some alcoholics can return to moderate drinking with no greater chance of relapse than if they abstained'. What angered some people was that the report, complete with the important conclusion just quoted, was released directly to the news media before the experts had been able to comment on it. Nevertheless, opponents of the idea of resumed normal drinking lost little time in making their views known. The National Council on Alcoholism, which was centrally involved in the formation of the Alcoholism Movement in the 1940s (see p. 45), hastily assembled two press conferences. At these the report was condemned as 'dangerous' and 'unscientific', even though no one at the NCA had had an opportunity to see a copy! The conclusion in respect of normal drinking was described as unethical, unprincipled and as 'playing Russian roulette with the lives of human beings'. There was even a suggestion that the entire report should have been suppressed.

Four years after the publication of the first Rand Report, the same team of researchers produced a further report of a four-year follow-up of the same treatment population.[6] Although it created less stir than its predecessor, the later report was based on firmer scientific foundations, because its authors were given the opportunity to correct some of the faults and answer some of the justified criticisms of the original. As before, however, the four-year follow-up report was subjected to considerable distortion in the media. A good example of this is an article appearing in the *International Herald Tribune* on 25 January 1980 with the heading: 'Study Says Alcoholics Can't Learn to Drink'; and subheaded: 'Reverses Findings of Four Years

Ago'. Any careful reading of the report itself will show how astonishingly unjustified these interpretations are, but this single example may help to show how irrational the reaction to normal drinking in former alcoholics often is.

The Sobell affair

The third major controversy over normal drinking in former alcoholics centred on the results of a well-known experimental trial of controlled-drinking treatment in California. Treatment aimed deliberately at controlled drinking rather than abstinence began to be developed and tested, once the results of Davies's and other follow-up studies became widely known. There was nothing in a learning-theory conception of alcoholism, favoured by the majority of psychologists working in the field, which ruled out the possibility of recovered control. Moreover, during the 1960s enthusiasm and optimism for behaviour-therapy solutions to all kinds of problems had reached a high level.

It was against this background that the young, husband-and-wife team of psychologists, Mark and Linda Sobell, became interested in trying out controlled-drinking treatment at Patton State Hospital, San Bernadino, California, in 1969. At the time, they were relatively inexperienced researchers. Mark was just completing a PhD in experimental psychology and Linda was still an undergraduate. But it is not true to suggest, as some media accounts have done, that they dreamt up the idea of controlled-drinking treatment out of their own heads. Apart from the extensive evidence of its occurrence following abstinence treatment and some previous Australian work on treatment aimed at moderation, the Sobells joined a team of experienced researchers who had conducted a number of preliminary studies leading up to the Sobells' own treatment trial.

The subjects were seventy male, 'gamma' alcoholics, that is those showing evidence of physical dependence on alcohol.

They were typically lower-class individuals with several previous hospitalizations for alcoholism and would be convention-
ally regarded as having a poor prognosis for any type of
treatment. Th▨e patients were initially selected for a con-
trolled-drin▨▨▨▨▨al by members of the hospital staff, principally on▨▨▨▨▨▨whether they had shown some control over
dri▨▨▨▨▨▨▨ad a supportive environment to return
▨▨▨▨▨▨oderate drinking as a goal. The forty
▨▨▨▨ration approach were equally and
▨▨▨▨ing called Individualized Behavi-
▨▨▨▨lled drinking as the treatment
▨▨▨▨conventional, abstinence-oriented
▨▨▨▨-C group).[7]
▨▨▨om treatment, all forty patients were
fol▨▨▨▨wo-year period, and extensive and detailed
infor▨▨▨▨ollected from the patients themselves and
from a▨▨▨wo relatives or friends. The Sobells' follow-up
procedur▨▨▨ere the most thorough and sophisticated assessment methods ever used in this area of research. The results
provided positive evidence in favour of the controlled-drinking-
treatment approach. At the end of the first year, the twenty
patients in the CD-E group were reported as functioning well
for an average of 70.5 per cent of days, compared with 35.2 per
cent of days for the nineteen subjects successfully contacted in
the abstinence-oriented CD-C group. This difference was maintained at the end of two years and, indeed, subjects in the CD-
E group were found to continue their improvement, relative to
the CD-C group, from the first to the second year, as shown by
the percentage of days spent in hospital or prison and by days
of abstinence. This is consistent with other evidence suggesting
that one important effect of behavioural treatment is to set in
motion a learning process which is continued through the
vicissitudes of the post-treatment period.

One fault of the Sobells' study was that the second-year
follow-up interviews were all conducted by Linda Sobell, who

was obviously not 'blind' to which treatment group the subject belonged. Partly in order to correct this potential source of bias, an additional third-year follow-up was conducted by an independent team of researchers headed by psychologist Glenn Caddy, in which all interviewers were kept blind as to treatment group. The results confirmed the superiority of CD-E to CD-C subjects on the measures of drinking and general adjustment employed by the Sobells.[8]

The integrity of the Sobells' research came under fire soon after the completion of the treatment trial. Irving M. Maltzman, who was then Chairman of the Department of Psychology at the University of California, Los Angeles, and a distinguished experimental psychologist, was asked to give an opinion on a proposal for research from someone unconnected with the Sobells but who wished to develop their work. Not being an authority on alcoholism, Maltzman sought the advice of Mary L. Pendery, a former graduate student of his who was director of an alcoholism-treatment programme in La Jolla, California. Together they visited Patton State Hospital, which the Sobells had recently left after completing the treatment phase of their research, where they encountered some anxiety on the part of hospital staff as to the welfare of the Sobells' controlled-drinking patients. Accordingly, in June 1973, they obtained permission from the hospital research committee to conduct an independent follow-up investigation.

After receiving financial support from various sources, Mary Pendery completed the main follow-up in 1979, although contact was maintained with subjects until 1981, representing altogether a ten-year follow-up of the Sobells' subjects. The next event of note was the circulation among the alcoholism-research community, during the earlier part of 1982, of a draft paper prepared by Pendery, Maltzman, and West. Although this paper contained important differences from the version eventually published, it led to a widespread rumour that the Sobells had falsified their results. Subsequently, a revised version appeared in the prestigious magazine *Science* on 9 July 1982.[9] Although this fell short of an explicit accusation of fraud, there were unambiguous assertions of serious scientific misconduct and, in an article in the *New York Times* dated 28 June 1982, Dr Maltzman is quoted as saying: 'Beyond any reasonable doubt, it's fraud.'

Media involvement in the affair intensified when it was made the subject of a special report in the CBS documentary programme *60 Minutes* on 6 March 1983. This programme is highly popular in the United States and its version of events was very unfavourable to the Sobells and their research. Many other reports and features appeared in American newspapers and magazines commenting on the controversy, again mostly to the detriment of the Sobells' reputation.

After the publication of the *Science* paper and some time before the *60 Minutes* programme had been broadcast, the Sobells had asked the President of the Addiction Research Foundation in Toronto, where they were now working, to set up a committee to investigate the charges against them. This committee was duly established and published its report in November 1982. The report became known by the name of the committee's Chairman, Bernard M. Dickens, Professor of Law, Criminology, and Community Health at the University of Toronto and an acknowledged expert on ethical issues in research. The three other committee members were all highly respected and distinguished academics who would have a great deal to lose by colluding with malpractice. None of the committee held appointments at the Addiction Research Foundation and none had any prejudices about controversies in the alcoholism field.

From the allegations in the *Science* article and the earlier draft, the *Dickens Report* isolated eight separate issues and discussed each in turn. We shall not have space to describe each of these issues individually but will concentrate on the two most prominent charges.[10]

1. One of the most devastating allegations was the assertion that thirteen of the subjects trained to do controlled drinking were rehospitalized in alcoholism units within approximately one year after discharge. Extracts from patients' records referring to these rehospitalizations formed the basis for a prominent table in both the *Science* article and the draft version; and by the use of the term 'new findings' in the title of the *Science* paper, it is clearly implied that the Sobells were either unaware of or deliberately suppressed these findings. Yet the committee were able to establish unequivocally that the Sobells had actually reported more rehospitalizations than the Pendery team. It concluded that the Sobells' published data were accurate and that there was no evidence of fraud or unethical behaviour.

2. The most sensational type of evidence produced by the Pendery team was that four of the subjects trained to drink in a safe manner had suffered alcohol-related deaths and these were vividly described in their papers. These deaths featured prominently in media accounts of the affair and, in the CBS *60 Minutes* programme, the presenter was filmed standing beside the grave of one of the dead patients. The conclusion most viewers would draw is that these patients had died as a direct result of receiving controlled-drinking treatment, despite the fact that the earliest death occurred six and a half years after treatment and the latest eleven years afterwards. However, in order to determine whether the incidence of deaths in the CD-E group was abnormally high, the committee looked at death rates for alcoholics in general and concluded that the four CD-E deaths, however tragic, were not unusual in number. What is more, in the CD-C abstinence-oriented control group, which was ignored by the Pendery team in their reports, at least six deaths had occurred within the same time period, of which it has been established that five were alcohol-related. In the light

of this, the committee found no reason to doubt the Sobells' integrity in the presentation of this part of the study.

The final verdict of the *Dickens Report* is as follows: 'The Committee's conclusion is clear and unequivocal. The Committee finds there to be no reasonable cause to doubt the scientific or personal integrity of either Dr Mark Sobell or Dr Linda Sobell.' This conclusion has subsequently been fully supported by a special investigator appointed by a Congressional Sub-Committee of the House of Representatives.

More recent findings

Following the Sobell affair, the dispute about the validity of the controlled drinking phenomenon has continued. One influential contribution has come from Edwards[11] who traced the seven allegedly normal drinkers originally described by D. L. Davies in 1962 (see pp. 90–4), extending the total follow-up of these individuals to between 29 and 34 years. The six subjects still alive were interviewed either in person or by telephone, relatives were also interviewed whenever possible, and GP and hospital records were inspected when available. The conclusion was that five of Davies's patients had experienced significant drinking problems, both during the follow-up period reported on by Davies and subsequently. The two remaining subjects were regarded as having engaged in problem-free drinking during the total follow-up period but one of these was said to have been never severely dependent on alcohol.

There are several points to make about these findings. First, taking Edwards' conclusions at face value, the results confirm that resumed normal drinking in alcoholics *can* happen, since it was shown that at least one of the men interviewed had become a problem-free drinker following a severe level of alcohol dependence. Secondly, material from retrospective interviews of this kind, especially over such a long time period, must be regarded as dubious. While it might be suggested that Davies's patients misled him by telling him what they thought he wanted to hear, the opposite kind of bias may have applied, to an unknown extent, to the information collected by Edwards. The best way to settle issues of this kind is in prospective studies in which definitions as to what is to count as 'normal' drinking or otherwise are decided on beforehand and not after the fact. Finally, however valuable a study it may have been, Edwards' follow-up is still only one among many sets of data that are relevant to the issue of controlled drinking; it is astonishiing that it should have been seized upon as the study which finally 'disproves' the existence of controlled drinking in former alcoholics.

In fact, there have been several recent studies whose results are much more favourable to the possibility of controlled drinking.[12] As one example, Nordström and Berglund,[13] two Swedish psychiatrists, followed up 70 male alcoholics showing 'good social adjustment' an average of 21 years after treatment. Defining social drinking as an absence of drinking problems in the year before follow-up and consumption of more than one drink per month, they found that social drinkers (21 subjects) were nearly twice as common as abstainers (11 subjects). Of the social drinkers, eleven had shown a relatively early change to social drinking via abstinence and ten a more gradual change. Five had experienced delirium tremens, indicating a very severe level of alcohol dependence before social drinking occurred. Another important finding was that temporary relapses during the follow-up period were less common among the social drinkers than the abstainers.

It is clear, then, that there exist variations, sometimes startling, in reported rates of controlled drinking by former alcoholics. These variations have been analysed by Stanton Peele.[14] He asserts that controlled drinking outcomes were common for a brief period in the mid to late 1970s; that by the early 1980s a consensus had emerged, at least in the USA, that very few severely dependent alcoholics could resume controlled drinking; and that this was followed by a further burst of studies reporting that return to controlled drinking was quite plausible and did not depend on initial severity of problems. He then attempts to demonstrate that cultural factors—national culture, ethnicity, subculture, and cultural era, as well as the professional and individual orientations of the scientists making the reports—are responsible for the wide variations in observed rates of controlled drinking.

While there may be much truth in Peele's arguments, there is a much more obvious factor contributing to the variations in question. This is simply *how* controlled drinking, or whatever term is preferred, is defined. An outstanding recent example of this is contained in an article by Helzer and his colleagues published in the prestigious *New England Journal of Medicine*.[15] In the Abstract of their article the authors write: 'The evolution to stable moderate drinking appears to be a rare outcome among alcoholics treated at medical or psychiatric facilities.' Again, this conclusion has been taken up enthusiastically by those who wish to argue that the phenomenon of controlled drinking is so rare as to be negligible. The percentages of surviving patients falling into each outcome category in this five- to seven-year follow-up of alcoholics who had received abstinence-oriented treatment is shown in Table 4.1. At first sight, the conclusion quoted above appears justified in that only 1.6 per cent of the follow-up sample were classified as 'moderate drinkers'. However, when it is realized how these moderate drinkers were defined, the authors' conclusion becomes much more puzzling. Any subject who had been abstinent for six or more months during the previous 36 months was excluded from the category and classified as an 'occasional drinker'. Thus, to be called a moderate drinker, a subject not only had to drink moderately but also, more or less, continuously. Why should it be insisted on that moderate drinkers should not be permitted to show any significant period of abstinence? In our own

Table 4.1. Drinking status of surviving alcoholics during three years before follow-up

Drinking status	Percentage
Abstainers	15.1
Occasional drinkers	4.6
Moderate drinkers	1.6
Heavy non-problem drinkers	12.2
Continued alcoholics	66.5

Adapted from Helzer, J. E., *et al.* (1985). The extent of long-term moderate drinking among alcoholics discharged from medical and psychiatric treatment facilities. *New England Journal of Medicine*, **312**, 1678–82.

research in Dundee,[16] we showed that one of the effects of controlled drinking treatment seems to be to increase the proportion of abstinence in the drinking pattern, presumably because the incentive value of drinking has been decreased in a new life-style. There is no reason why the same should not also apply to moderate drinking arrived at without professional help. More relevant than the number of months drinking or abstinent is whether or not subjects perceived themselves as moderate drinkers who occasionally preferred periods of abstinence, for whatever reason, or whether as abstainers who had fairly frequent but unharmful slips. Although no data on this are presented by Helzer and his colleagues, the former appears more likely. In any event, if the occasional drinkers in Table 4.1 are added to the moderate drinkers, to form a group of 'controlled drinkers', their frequency immediately climbs to 7.2 per cent, a figure of much more clinical significance.

But that is not all. There is also a group in Table 4.1 called 'heavy non-problem drinkers'. These were subjects who exceeded the criteria for moderate or occasional drinking (no excessive drinking, defined as seven or more drinks per day on four or more days in any one month in 36) but who showed no evidence of any social, medical, or legal problems from drinking throughout the entire three-year period before assessment. This absence of problems was confirmed by interviews with relatives and thorough inspection of official records of various kinds. (Incidentally, all reports of moderate drinking in this study were verified by collateral information whereas reports of total abstinence were completely unsupported by independent data.)

The question here is why these subjects, whose only apparent transgression was to drink in excess of the author's arbitrary definition of moderate drinking, were excluded from a consideration of the frequency of controlled drinking outcomes. Since, in a thorough investigation, they had shown no evidence of drinking problems over a protracted period of time and since the overriding aim of treatment must surely be to improve the quality of peoples' lives by eliminating the alcohol-related problems they had once experienced, the conclusion must be that this was done on ideological rather than scientific grounds. It might be objected here that these 'heavy' drinking non-problem subjects were consuming alcohol at levels likely to increase the risk of medical damage, in terms of public health criteria for 'safe' drinking. But this a separate issue, since large numbers of people in our society, who do not experience any obvious problems with drinking and certainly do not seek treatment for such problems, also exceed these limits. At the very least, it seems extraordinary that this group of subjects should be ignored in the authors' summary of an article which has been widely quoted as centrally relevant to the controlled drinking controversy. If the occasional drinkers, moderate drinkers, and heavy non-problem drinkers in Table 4.1 are added together to form a group of 'non-problem drinkers', then their percentage (18.4) would exceed that of the unconfirmed abstainers (15.1).

There are several other examples of the apparent double standard in definitions of controlled drinking and abstinence that has been illustrated above.[17] The more general point, however, is that recent evidence, particularly long-term follow-ups of the results of abstinence-oriented treatment, strengthens rather than weakens the case for the existence and clinical significance of controlled-drinking outcomes. Nevertheless, there is a difference between research evidence of this kind and practical applications in treatment. Certainly, we do not yet know enough about the recovery process to advocate the controlled drinking goal for *all* problem drinkers. It is important to be clear about this. Our view is that, at the present stage of knowledge, most severely dependent problem drinkers should be advised to aim for abstinence rather than controlled drinking.

We have stressed this point because it is possible that this book may be read by people who themselves have drinking problems and the material in it may have personal significance for them. The evidence on resumed normal drinking, particularly the statement that there appears to be no upper limit of dependence which rules out the possibility of control, might be construed by some as an invitation to attempt a return to normal drinking themselves. Nothing could be further from our intentions. *There are many serious problem drinkers whom we would strongly advise to opt for total abstinence and not to attempt controlled drinking.*

The difficulty here is that, when one speaks of normal drinking in alcoholics, it is often assumed that one is describing an 'easy option' compared with abstinence; problem drinkers may believe that a return to normal drinking is something they can accomplish without much effort or pain. The truth is the very opposite. For those with a severe level of dependence and a long history of heavy drinking, a harmfree pattern of use is extraordinarily difficult to achieve and this is why such persons should typically be advised to abstain. *A decision to aim for controlled drinking in someone with a serious and long-standing problem should only be made after seeking competent professional advice.*

Summary

Since the topic of normal drinking in former alcoholics is of such great importance, not only for theory but practical treatment, it will be advisable to include a separate summary section devoted to the current state of knowledge about it.

1. The contention that persons with previous drinking problems who have recovered the ability to drink normally were not 'real' or 'true' alcoholics in the first place has been shown to be false. The evidence indicates that there is no upper limit to severity of physical dependence which absolutely precludes the possibility of recovered control. There *is* evidence that normal drinking becomes increasingly rare as severity of dependence increases.

2. Enough studies have now been conducted using objective, predetermined criteria and corroborative sources of information to guarantee the validity of resumed normal drinking. There remain several important research questions regarding the quantity and quality of the normal drinking engaged in.

3. Early evidence that normal drinking can be maintained over extended periods of time without relapse has been confirmed many times over. The Rand data show that severity of dependence is not the only major determinant of the relative relapse rates of normal drinking and abstinence; age, marital, and employment status are also important.

4. The description of resumed normal drinking as a rare event of no theoretical or clinical significance cannot be justified. Compared with abstinence, normal drinkers have been

found to represent substantial minorities in several studies and to be in a majority by others. Several factors affect the relative proportions of normal drinkers and abstainers, most notably the prior degree of dependence. However, socio-cultural factors and the beliefs the individual has been exposed to, as shown particularly by the effects of attendance at AA meetings, are also relevant.

5. At the present state of knowledge of the recovery process, severely dependent problem drinkers should normally be advised to abstain rather than aim for controlled drinking. This is because a controlled drinking goal is extraordinarily difficult to achieve in severely dependent individuals and, except in special circumstances, total abstinence offers the best chances for success.

6. There is now a consensus that the most important application of the controlled drinking goal and its associated methods of behaviour-change is in the treatment of low or moderate dependence problem drinkers and in the prevention of more serious harm in those with recent problems. The fact that such problem drinkers will not be deterred from seeking or entering into treatment by the prospect of total and lifelong abstinence increases interest in the potential of controlled-drinking methods in this area. The application of the controlled drinking goal in community-based interventions will be described in more detail in Chapter 9.

Household survey research

The evidence on resumed normal drinking shows that alcoholism is not the irreversible condition it was portrayed as being in the disease theory. Substantially the same conclusion emerges from a quite different source of evidence, which provides in addition other valuable insights into the nature of problem drinking in the real world. Rather than studying samples of hospitalized or otherwise diagnosed 'alcoholics', the evidence to be considered in this section ignores institutionalized populations and concentrates instead on 'normal' (non-institutionalized) samples of alcohol consumers.

Ironically, this line of research acquired its initial impetus from the growing acceptance of the disease theory in the early 1960s. When alcoholism began to be recognized as a serious national problem, issues relating to the provision of resources to deal with it started to be discussed, and politicians and policy-makers began to demand answers to the question, 'How many alcoholics are there?' This was the province of 'epidemiology', the study of the prevalence, incidence, and social ecology of specific diseases. Early attempts to answer the question during the 1950s by indirect estimation, such as Jellinek's suggestion for using rates of cirrhosis mortality, had been shown to be inadequate. Moreover, there was a presumption in the Alcoholism Movement that the alcoholics seen in treatment represented only the tip of the iceberg; the number of 'hidden alcoholics' was unknown but universally assumed to be large. Thus, in the effort to answer the insistent question of numbers, researchers were driven out into the streets and forced to knock on the doors of the public at large. The irony is that, in using this direct approach to the prevalence of the disease, this research provided some of the strongest grounds for rejecting the very concept on which it was founded.

In this section we shall concentrate on the most famous household survey in the field, which yields all the major lessons to be drawn from this kind of research for an evaluation of the disease theory. This is the survey directed by Don Cahalan and his associates from the Social Research Group (now the Alcohol Research Group), University of Berkeley, California. Cahalan's survey employed a random and representative sample of 1359 adult residents of households throughout the

United States interviewed initially in 1964–5 and re-interviewed three years later.[18]

The first lesson is that problem drinking is much more common in the general population than is implied by a disease concept that emphasizes the predisposition to alcoholism of a relatively small group of individuals. Cahalan put questions about eleven specific problem areas connected with drinking and respondents were asked whether they had experienced these problems within the last three years. By various combinations of scores from Cahalan's data, the 'prevalence' of problem drinking in the general population could be made to vary enormously and to include, under plausible-sounding criteria, roughly one third of the adult male population. Thus 43 per cent of men and 21 per cent of women admitted to having had at least one problem during the past three years. It would obviously be absurd to call all these people 'problem drinkers', but no natural cutting-off point could be found to distinguish those with problems from those without. Nevertheless, admitting that it was an essentially arbitrary decision, Cahalan used a score of over seven on an index of current problems, which took into account the frequency and severity with which each problem was experienced, to demarcate a category of problem drinkers with fairly severe current problems. Under this definition, 9 per cent of the total sample, comprising 15 per cent of men and 4 per cent of women, were classified as problem drinkers. This was equivalent to about nine million of the adult American population.

If all these people were the 'hidden alcoholics' of the disease-theory assumption, then there were obviously an awful lot of them! It eventually became clear, however, that these problem drinkers differed in systematic ways from the clinic alcoholics on which descriptions of problem drinking had, up to then, been based. Firstly, clinic alcoholics are much more likely to be unemployed or in a marginal job and to be divorced or separated than survey-identified problem drinkers, suggesting that alcoholism treatment provides one disposition for the 'spare' people in our society. Secondly, the problem drinkers are much younger on average than the clinic alcoholics; the highest concentration of male problem drinkers is in the early 20s, whereas clinic alcoholics are typically in their 40s or 50s. These and other differences have led epidemiological researchers to speak of 'the two worlds of alcohol problems'. The earlier concentration on relatively deteriorated alcoholics found in treatment centres and penal institutions, and the corresponding ignorance about the nature of drinking problems as they are manifested in the world outside the institutional setting, had resulted in a selective and misleading picture.

Perhaps the most important finding to emerge from household-survey research is that drinking problems in the natural environment are much more volatile than would be expected on the basis of any disease concept. For example, in addition to the 9 per cent in Cahalan's study with fairly severe current drinking problems, there were a further 9 per cent who said they had experienced such problems in the past but were no longer troubled by them. Many of the young men who admitted to problems at the first interview were found to have 'matured out' of them by the second. Cahalan's results showed, in summary, that 22 per cent of men and 9 per cent of women had changed their drinking problem status within the short three-year period covered by the survey. Thus, fluctuations in problem drinking, both into and out of problem status, are much more common than is suggested by any concept of alcoholism involving an unavoidable deterioration if drinking is continued.

The same kind of conclusion can be reached from an analysis of the relationships between specific problems. Research by Clark and Cahalan published in 1976[19] looked at some longitudinal data from this point of view. Their sample consisted of 615 white males aged between 21 and 59 living in San Francisco at the time of a first interview in late 1967. These men were re-interviewed in early 1972, the gap between interviews being roughly four years. The results showed that the presence of problems at the first interview was a poor predictor of problems at the second. It is true that continued involvement in some alcohol problems was the rule rather than the exception for those with problems in 1967 but, even here, a substantial proportion reported a complete absence of problems in 1972. Moreover, the presence of a particular problem at first interview was not even a good predictor of the presence of the same problem on the second occasion. All this evidence is quite out of keeping with a model which suggests a progression from less to more severe problems and a gradual accumulation of more and more problems as the condition worsens.

Returning to Cahalan's national survey data, it is also of interest to examine the reasons subjects gave for having changed their drinking habits. Those who had increased their drinking, and had higher problem scores at the second interview, gave as their principal reasons greater financial ability to purchase alcohol, the influence of other people, and having more opportunities or time for drinking. Few said that they were drinking more because of increased tensions, stresses, or other general problems in living. Similarly, those who were drinking less and had lower problem scores emphasized financial reasons, increased responsibilities, having less need or desire for alcohol, becoming older and more mature, and reasons of health. Mention of moral and religious reasons or feelings of guilt were much less frequent. Thus, reasons given for both increases and decreases in drinking emphasized environmental and role factors rather than the psychological factors which would be predicted as important by some disease concepts.

The evidence in this section all goes to demonstrate that drinking problems in the natural environment do not show the fairly inflexible course described in Jellinek's classic analysis of the questionnaire responses of AA affiliates (see p. 79). It might still be argued, however, that, among the so-called chronic alcoholics found in formal treatment programmes, such an invariable course would be found leading up to their present deteriorated state. There has been a substantial amount of research on this topic and, although a digression from household surveys, it will be convenient to consider it here.

Briefly, the evidence shows that the behavioural and social consequences of excessive drinking, which featured prominently in the phases of addiction described by Jellinek, do not occur in any predictable order and may be found at virtually any stage in the sequence of events in a problem drinker's life. There has recently been a suggestion that, if attention is confined to the core symptoms of dependence, some degree of order may be shown, in that an increasing need for alcohol and the increased importance of drinking in a person's life usually precede the onset of symptoms of withdrawal.[20] But this is not Jellinek's order, since 'loss of control' appears early in the sequence and not late. Moreover, some patients show considerable idiosyncratic variation from the typical pattern. Finally, the demonstration of some kind of order in a retrospective analysis has no bearing on the likelihood that early symptoms will lead on to later symptoms. Indeed, the household-survey research summarized in this section shows clearly that there is no inevitability about the progression of these 'symptoms' of alcoholism.

Spontaneous remission

Household surveys have also made a major contribution to another research tradition which undermines the notion of problem drinking as a stable and lasting phenomenon. This is the study of 'spontaneous remission'—a medically derived term referring to improvements in a disease condition without the intervention of formal treatment. The basic data in this line of research are mainly the same results of community and national surveys considered above, with the addition of a few studies of people who are selected for treatment but who, for one reason or another, do not receive it. However, the implications of looking at these data from the point of view of spontaneous remission are rather different and concern a further element of disease-theory assumptions, the need for treatment if further deterioration is to be avoided.

Some of the most telling evidence of resumed normal drinking, which was covered earlier in this chapter, comes from studies of spontaneous remission. For example, Kendell's research at the Maudsley Hospital (see p. 93) was done on untreated alcoholics. Other relevant research strongly suggests that alcoholism tends to disappear with increasing age and that a large proportion of this disappearance is due to spontaneous recovery. And we have already seen that many of the problem drinkers identified in Cahalan's survey tended to 'mature out' of their problems without receiving any professional help. In a comprehensive review of the effects of alcoholism treatment,[21] it has been calculated that roughly 15 per cent of untreated alcoholics spontaneously achieve at least six months of total abstinence and a further 40 per cent show improvements in their drinking.

As with the correlates of changes in drinking-problem status and the accompaniments of resumed normal drinking, the main determinants of spontaneous remission appear to be changes in life circumstances involving marriage, employment, health, and finance. However, one recent study suggests that the reasons people give for stopping or cutting down drinking may bear little relevance to reality.[22] Some American studies report a very large percentage attributing their change to some kind of religious experience. In Britain, on the other hand, more mundane reasons tend to be given, including the changes in life circumstances mentioned above. But if problem drinkers are followed up, as opposed to being asked in retrospect why they stopped or changed, 'life events', such as changes in job and family circumstances, are no more common before times when people gave up or cut down than before times when they drank more or stayed the same. If the same people are asked in retrospect why they changed, they commonly attribute their change to such events, even though similar events have occurred at times when they did not change their drinking. It is possible that the same kind of thing was happening in the American studies; given the prevailing moral climate described in Chapter 2 (pp. 58–61), drinkers made sense of their change in drinking by attributing it to divine forces.

Laboratory investigations of drinking and intoxication

Both the evidence from follow-up studies of treated problem drinkers and the evidence from household surveys of the general population have undermined the assumption of the irreversiblity of alcohol problems, even for the most severe variety. This in turn has made dubious the central theoretical construct of the disease theory of alcoholism, the notion of an irreversible 'loss of control' over drinking, as well as casting doubt on the disease-theory version of 'craving'. However, there is another body of evidence which is more directly relevant to these core constructs, and which is even more damaging to their credibility. This comes from the strictly 'experimental' setting of observations of drinking behaviour made under controlled conditions in the laboratory.

Research of the Mendelson laboratory

Up to the 1960s the main evidence on which formulations of loss of control and craving had been based consisted of the clinical observations of psychiatrists and other professionals and the retrospective reports of their experiences by AA members. Thus, the original source of this evidence was, in both cases, the personal recollections of alcoholics. However well-intentioned and devoted to accuracy such reports may be, they are clearly suspect for scientific purposes for several reasons. They may be influenced by alcohol-induced amnesias, the general fallibility of human memory, conscious or unconscious anticipation of what the interviewer wants to hear, the biasing effects of beliefs and theories the alcoholic or the interviewer may subscribe to, and a host of other factors. The point is that no systematic observations of alcoholics' actual drinking had been made under laboratory conditions and subject to accepted rules governing the collection of scientific data.

This all-important gap in our knowledge was filled by a pioneering programme of research at the National Institute of Mental Health, Maryland, under the direction of J. H. Mendelson. Besides the introduction of scientific methods, Mendelson's research made another significant innovation by giving alcohol to alcoholics for research purposes, which had been taboo up till then. The consequence was that, for the first time in the history of the study of alcoholism, speculation and dogma were forced to confront objective and systematic observations of the alcoholic's drinking behaviour.

Since there may be ethical objections to these experiments, it is important to note that all subjects were volunteers from a nearby correctional institute, all lived in individual bedrooms with access to a day-room, games, and other recreational activities, all were offered psychotherapy and were provided with close medical supervision of their health during the experiments, the purpose of which was carefully explained to them. Thus, every effort was made to ensure that they experienced no lasting ill-effects from their participation. The subjects were mostly homeless men with repeated imprisonments for drunkenness over a number of years and all showed clear evidence of physical dependence.

The design of Mendelson's experiments was based on the

'operant methodology' derived from the radical behaviourism of B. F. Skinner. The main principles of operant learning, or instrumental learning as it is also called, and the relevance of these principles to drinking behaviour, will be described in Chapter 7 (pp. 220–36). Here we may merely note that this behaviourist methodology is not at all concerned with arriving at generalized conclusions regarding differences on average measures between defined groups of individuals, as is the aim in conventional psychological research. Instead, the object is to discover what enviromental factors are associated with the occurrence of specified behavioural responses, such as excessive drinking, in the individual case. This is done by attempting to bring the behaviour under control through the manipulation of environmental rewards and punishments.

The first experiment to use operant methodology in this fashion was reported by Mello and Mendelson in 1965.[23] Two alcoholic subjects had free access to an operant-conditioning apparatus, from which they could earn drinks by pressing a translucent key in the presence of an appropriate visual stimulus. Subjects could work the apparatus at any time during day or night over periods of eleven and fourteen days respectively. The immediate finding was that, despite the fact that their performance at the apparatus was highly inefficient, subjects were able to maintain high levels of blood-alcohol concentration (BACs) ranging from 150 mg per 100 ml to 250 mg per 100 ml, or from about two to three times the legal driving limit in Britain. Owing to their high tolerance for alcohol, they did not become grossly drunk and displayed only mild intoxication with alteration to mood but without marked impairment to speech or movement.

In later work, the Mendelson team became less interested in studying spontaneous drinking patterns as such and more concerned with assessing the contribution of various factors which were hypothesized to affect alcoholic drinking. In one experiment,[24] the emphasis was on the 'cost' of alcohol, cost being defined in the laboratory situation as the amount of work required to obtain it. The basic finding was that subjects required to complete sixteen consecutive correct responses to earn an alcohol reward showed average BACs roughly twice as high as subjects who had to complete thirty-two consecutive correct responses. In other words, the amount of alcohol consumed was a predictable consequence of the degree of effort required to earn it.

Having given a flavour of the kind of research done by the Mendelson team, we will now summarize its main implications for the concepts of loss of control and craving. First, and in a way which is most obviously in contradiction of 'loss of control', none of the subjects studied, who all conformed to classic diagnostic criteria for the presence of alcohol addiction and who were all placed in a situation where they could determine the amount and pattern of their own drinking, tried to drink themselves into a state of oblivion. This is not to claim that these men had never engaged in this kind of behaviour in the natural environment, but it is to assert that drinking to gross intoxication is not an inevitable feature of alcoholic drinking. Second, no subject chose to drink all the alcohol available to him, even when no effort at all was required to obtain it. Third, these alcoholics showed positive sources of control over their drinking in the following ways: they drank to maintain high but roughly constant BACs during shorter drinking sessions; they did not drink continuously but spontaneously started and stopped drinking sessions over longer experimental periods; they tended to work for and drink moderate amounts of alcohol and did not drink it as soon as it became available; some subjects chose to taper off their drinking in order to avoid or reduce the withdrawal symptoms which they knew would occur at the end of the experiment; other subjects chose to work over one- or two-day periods to accumulate alcohol rather than drink to abolish partial withdrawal symptoms as soon as they occurred. All these observations are inconsistent with the concept of loss of control in the sense of an inability to stop once drinking has begun. They are also inconsistent with the related idea of craving in the sense of an uncontrollable urge to consume more and more alcohol during a drinking session.

Other experimental investigations of alcoholics' drinking

Following the precedent set by Mendelson and his colleagues of giving alcohol to alcoholics for research purposes, there have appeared a great many reports of experimental investigations of alcoholics' drinking using operant methods. These studies have demonstrated in various ways that, within a laboratory or a controlled hospital in-patient setting, the drinking of chronic alcoholics is a predictable function of environmental contingencies. It will clearly be impossible even to mention all these studies here and we shall confine ourselves to one highly ingenious experiment conducted at the John Hopkins Hospital, Baltimore, under the direction of Miriam Cohen.[25]

The focus of this study was on cost factors in alcoholic drinking but, in contrast to the Mendelson research, the target behaviour was abstinence. The broad aim was to discover what rewards were necessary to 'buy' abstinence from alcoholics. This was done by examining the interrelationships between cost factors and two other, highly relevant variables—a 'priming dose' of alcohol and a delay in the occurrence of the reward for abstinence. The subjects were four chronic alcoholics who had all lost jobs and been hospitalized for alcoholism in the past. Selection and living arrangements were similar to those described for the Mendelson experiments.

Subjects were allowed to purchase a relatively large amount of alcohol every third day of the experiment. On subsequent days, each was offered a certain amount of money to abstain for an entire day. If the subject did not abstain, the financial incentive for abstinence was increased on the next occasion; if he did abstain, it was lowered. Subjects were not informed how payments for abstinence were calculated. The results showed that abstinence could be bought from each subject for a varying amount of money ranging from seven to twenty-seven dollars and this, in itself, is incompatible with a simplistic notion of loss of control.

However, it is the effects of the other variables which are of chief interest here. Figure 4.1 shows what happened when subjects were given a fairly large priming dose of alcohol on the morning of the day on which payment for abstinence was offered. A varying amount of alcohol in the priming dose disrupted abstinence for each subject and the reward which had previously been effective in buying it now proved insufficient. Increasing the magnitude of the reward, though, did eventually succeed in establishing abstinence, as shown by the results from two subjects displayed in Fig. 4.1. The same kind of picture emerged from the delay-of-reinforcement condition. Increasing the delay before money was given to the subject initially disrupted abstinence, but increasing the amount offered eventually reinstated it.

The results of this experiment suggest how a belief in the concept of loss of control could have arisen, while at the same time showing it to be invalid. It is consistent with this idea that, having consumed a certain amount of alcohol to prime them, chronic alcoholics will reject abstinence even though it would appear to be clearly 'rewarding' for them and even though they had accepted it previously when sober. The crucial point, though, is that the drinking of these men could be brought under control once more by providing a sufficiently large incentive. Thus, any first impression of an uncontrollable craving for alcohol would have been an illusion. Delay of reward is also highly relevant to alcoholic drinking, since it may be assumed that the rewards for abstinence in the real world are usually more remote than the immediate gratification expected from drinking. But again, it was shown that alcoholics

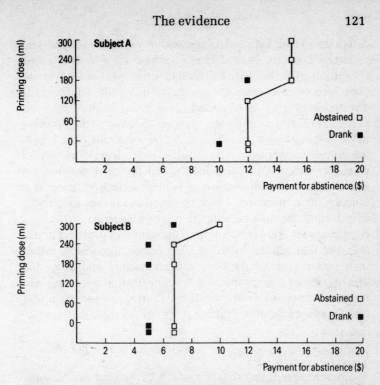

Fig. 4.1. Relationships between amount of priming dose, magnitude of reward, and reinstatement of abstinence [reproduced with permission from Cohen, M., Liebson, I. A., Faillace, L. A., and Speers, W. (1971). Alcoholism: controlled drinking and incentives for abstinence. *Psychological Reports*, **28**, 575–80].

will tolerate a delay of up to twenty-one days in payment for abstinence, provided the incentive is sufficiently inviting.

At this point, one crucial aspect of the evidence from laboratory experiments that goes against the traditional position must be singled out for emphasis. This is the general finding that the drinking behaviour of alcoholics has been shown to be *operant* behaviour. That is to say, in the classic definition of the term, it is behaviour which is 'shaped and maintained by its environmental consequences'. The broad conclusion from this is that alcoholics' drinking is subject to the same kind of laws as govern normal drinking or, indeed, goal-directed behaviour of any kind. This is not to deny that the particular reinforcement contingencies applying to the drinking behaviour of alcoholics will show differences from those applying to non-alcoholics; this is self-evident from the fact of markedly different drinking habits. But this does not obscure the demonstration that drinking behaviour in both cases is a function of environmental contingencies and is thus modifiable in principle by manipulating those contingencies. This is a completely separate conception of the nature of problem drinking from that offered by the disease perspective which, rather than pointing to similarities between normal and abnormal drinking behaviour, proposes a sharp distinction between them in the concepts of loss of control and craving, and assumes that one obeys a qualitatively different set of laws from the other.

Priming-dose experiments

A different line of investigation which has proved troublesome for the disease theory was prompted by observations reported in 1966 by psychiatrist Julius Merry, who entitled his article 'the "loss of control" myth'. In this much-quoted experiment,[26] nine in-patients, diagnosed as gamma alcoholics with loss of control, were given one fluid ounce of vodka disguised in orange juice and presented to them as a vitamin drink at breakfast. (It had previously been established that the vodka in the mixture was undetectable). In the control condition a similar proportion of water replaced vodka in the mixture. Subjects were given one or the other mixture in two-day sequences over sixteen days and during the late morning of each day were asked to record any degree of craving experienced on a simple five-point scale by nursing staff, who were unaware of the purpose of the experiment and of the fact that alcohol was being secretly administered to patients. The results showed no difference in average craving scores between the alcoholic and the non-alcoholic mixture. There was also no statistical difference in the number of occasions when *any* craving was experienced; indeed, there were slightly more such occasions following the non-alcoholic mixture.

Merry interpreted these results as refuting the hypothesis that small quantities of alcohol trigger off a biochemical reaction leading to craving in gamma alcoholics. He became interested in this topic after observing that his relapsed patients had usually omitted to take their prescribed Antabuse (a drug that leads to unpleasant feelings of illness if alcohol is taken) two or three days before they embarked on a binge. This suggests the existence, says Merry, of a 'mental set' to drink preceding the first ingestion. For this reason he concluded that psychological and environmental factors play a more important role in the loss of control reaction than was commonly supposed.

Despite the deserved amount of attention attracted by Merry's pioneering experiment, it can be criticized for employing crude measures of craving and on other technical grounds. However, there then followed a series of experiments, which we shall not be able to describe fully here,[27] that were based on Merry's design but which employed more sophisticated measuring techniques and also made the important distinction between the physiological and the psychological determinants of craving and loss of control. The design used in these experiments is known as the 'balanced placebo design' (see also Chapter 8, pp. 257–8). Subjects are either given alcohol or a soft drink and either told they are drinking alcohol or told they are drinking a soft drink. This results in four groups of subjects in the two-by-two classification shown in Fig. 4.2.

This was the design used in an experiment by G. Alan Marlatt and his colleagues from the University of Washington, Seattle.[28] One difference from Merry's experiment was that it

	Told soft drink	Told alcohol
Given soft drink	10·94	23·87
Given alcohol	10·25	22·13

Fig. 4.2. Average amounts of alcohol (fluid ounces) consumed by groups in a balanced placebo design [adapted from Marlatt, G. A., Demming, B., and Reid, J. B. (1973). Loss of control drinking in alcoholics: an experimental analogue. *Journal of Abnormal Psychology*, **81**, 233–41].

was actual drinking behaviour which was the focus of interest rather than just craving for alcohol. The subjects, who all had long-standing problems with alcohol but who were not currently hospitalized or seeking treatment, were informed that they were taking part in a 'taste-rating task', the object of which was to discover whether they could discriminate between three allegedly different brands of vodka or tonic, as the case might be. Twenty minutes after receiving the priming dose consistent with the experimental group shown in Fig. 4.2, subjects were seen individually and were told that they could drink as much as they liked of the beverage in front of them in order to come to a decision about taste and were then left alone for fifteen minutes. During this period, they were secretly observed by one of the experimenters who did not know which group the subject had been placed in. As well as thirty-two problem drinkers, a sample of thirty-two non-alcoholic, social drinkers was also studied.

Averages of the total amounts consumed by probem drinkers in the different groups are shown in Fig. 4.2. The overall results showed that the problem drinkers drank at higher levels than the social drinkers and that, for both types of subject, the main determinant of the amount drunk was instructional set and not the true contents of the drink. The authors conclude that their findings provide strong evidence of the important role of cognitive factors in the occurrence of loss-of-control drinking, but are quick to point out that they have provided only a laboratory analogue of loss of control in this experiment. Despite this qualification, the findings of priming-dose experiments provide strong support for an 'expectancy' interpretation of excessive drinking by alcoholics.

The possibility of a physiologically based craving has been revived by the results of recent experiments by Ray Hodgson and his colleagues from the Addiction Research Centre, Institute of Psychiatry, London.[29] The Hodgson group had certain technical criticisms to make of the priming-dose experiments described above and objected also to the artificiality of the situations in which drinking behaviour and experiences were recorded. Their main point was that neither of the earlier experiments had distinguished between different levels of severity of dependence among the subjects used. Owing to the earlier delineation by Griffith Edwards of the alcohol-dependence syndrome and the development at the Addiction Research Centre of a questionnaire to measure severity of dependence, the Hodgson experiments were able to divide subjects into those who showed mild or moderate dependence and those who showed severe dependence.

Employing the balanced placebo design and relying chiefly on speed of drinking as a measure of craving, these researchers were able to demonstrate that physiologically mediated processes (the bodily effects of alcohol) assumed a greater importance among severely dependent subjects than psychological factors (what the subjects had been led to expect) in the production of craving. For the mildly to moderately dependent group, the results were similar to those found by Marlatt. However, even in the severely dependent, the influence of expectations was still apparent and was combined with physiological factors in its effects on drinking. From the point of view adopted by the Hodgson group, these results are quite consistent with a learning-theory interpretation of craving, as we shall see in Chapter 7.

Summary

To summarize the conclusions which may be drawn from laboratory investigations of alcoholics' drinking, it must be recalled that, since the early nineteenth century, the twin concepts of loss of control and craving have formed the cornerstone of the disease theory of alcoholism. They have been especially important as justifications for total abstinence as the goal of treatment and for the attempt to absolve the alcoholic from responsibility for his actions.

The general conclusion from the evidence briefly reviewed is that, certainly in their present form, the concepts of loss of control and craving are no longer useful in the understanding and treatment of problem drinking and should be abandoned. As descriptions of alcoholics' behaviour and experiences, they are patently inaccurate; as explanations for the fact that alcoholics drink more heavily than other people, they are either logically circular and therefore untestable or, to the extent that they *are* testable, clearly contradicted by the evidence. In the limited sense of physical dependence based on the occurrence of withdrawal symptoms, 'loss of control' may have some role to play in the maintenance of alcoholic drinking, but here loss

of control can be seen to be reversible after periods of abstinence. Similarly, in the restricted senses of the alcoholic's
painful awareness of minor withdrawal symptoms while drinking and the conditioned association of these withdrawal symptoms with internal and environmental cues (see Chapter 7), the
concept of craving also has some validity. In both cases,
however, craving is not irreversible.

The distribution of alcohol consumption

All the evidence so far considered in this chapter has come
from studies of the behaviour of individuals—from follow-ups
of treated problem drinkers, household surveys of drinking
practices, and experimental investigations of drinking and
intoxication. There is also an important source of evidence,
equally embarrassing for the disease perspective, which derives
from the aggregate behaviour of large numbers of people. This
research is concerned with the way the consumption of alcohol
is distributed in a society—that is, with the relative numbers
and percentages of people who consume alcohol in various
quantities.

The first scientist to become interested in the distribution of
consumption was the French mathematician and demographer
Sully Ledermann who wrote chiefly in the 1950s.[30] Ledermann
reasoned that per capita alcohol consumption in a homogene-
ous population was unlikely to follow a 'normal distribution'—
the symmetrical, bell-shaped curve which describes the distribution of a great many naturally occurring phenomena, like
height. This was because it was impossible to drink less than a
zero quantity and, therefore, if consumption were normally
distributed, there would be an upper limit to quantities consumed of roughly twice the average level in the population. But
it was well known for quantities well in excess of twice the
average to occur. Rather, Ledermann hypothesized, alcohol

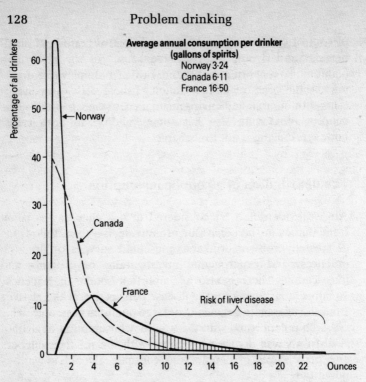

Fig. 4.3. Examples of Ledermann (log-normal) distributions of alcohol consumption [reproduced with permission from Kalant, O. J. (1971). *Drugs, society and personal choice*. Addiction Research Foundation, Toronto].

consumption would follow a 'log-normal' distribution, examples of which are given in Fig. 4.3. In this distribution, the largest number of consumers drink nothing or a relatively small amount, a substantially smaller number drink somewhat larger amounts, and a very small number drink very large amounts. Thus, the distribution is not symmetrical but 'positively skewed', with the highest point at the left-hand side and a long, progressively thinner right-hand tail (see Fig. 4.3). Rather than height, the distribution of alcohol consumption is more akin to that of personal income in society.

Ledermann supported his ideas with sets of data from a remarkable diversity of sources, ranging from a study of alcohol consumption among 1750 motorists in Chicago in 1935–7, through the blood-alcohol concentrations of 624 individuals attending a social-security centre in France, and the consumption of wine by 181 French tuberculosis patients before their illness, to the quantities of spirits bought by 28 000 Swedes in 1948. All these sources were used to support Ledermann's main thesis that the distribution of alcohol consumption was lognormally distributed.

To make his model more useful for predicting the extent of alcohol problems, Ledermann added a further crucial assumption. This was that the dispersion, or degree of variation about the average, of consumption levels was constant across different drinking populations. In other words, irrespective of the average level of consumption, distributions from different countries or other populations would be spread out to the same degree. Thus the only way distributions could differ among themselves was in the average level and this led to the Ledermann hypothesis being described as a 'single-parameter model'. The reason Ledermann gave for assuming the dispersion of consumption distributions to be constant was that there was a biologically fixed upper limit on the amount it was possible to drink, which he estimated at one litre of pure alcohol a day.

Two important consequences follow from this. First, since all distributions have a similar dispersion, there is a fixed relationship between the average of a distribution and the number of people drinking over a certain amount—for example, the amount at which liver damage or some other alcohol-related problem is likely to occur—and all that is needed to predict the number of problem drinkers in a population is a knowledge of the average level of consumption. This is shown in Fig. 4.3, which gives distributions from three countries; the higher the average consumption of the countries shown, the greater the proportion of the curve in the hatched area and the greater the number at risk of liver disease. The second consequence is that changes in average consumption over time within a population will lead to predictable changes in the number of those at risk for problems. In particular, an increase in average consumption will lead to a disproportionate increase in the number of alcohol problems.

Ledermann wrote almost entirely in French and his ideas were virtually unknown at the time to British and American scientists. They might well have remained obscure in the English-speaking world but for a short article which appeared in the *Quarterly Journal of Studies on Alcohol* in 1968 by Jan de Lint and Wolfgang Schmidt,[31] two epidemiological researchers from the Addiction Research Foundation in Toronto. This article reported the results of a survey conducted in the early 1960s into the purchase of distilled beverages and wines from retail stores in Ontario. Fortunately for researchers, all such purchases in Ontario must be recorded and customers are required to fill in their names and addresses and amounts ordered, thus providing a very convenient source of data.

De Lint and Schmidt's main conclusion was that the Ledermann log-normal curve gave a remarkably good fit to the actual distribution of quantities bought by customers over a one-month period. They also speculated as to why a simple logarithmic relationship should provide such an excellent approximation to what appeared to be fairly complex behaviour—why the drinking of all members of a population seemed to be influenced in some mysterious way by the level of average drinking habits. Ledermann himself had hypothesized that the particular shape his curve took was due to a 'snowball' effect of drinking which assumes that individuals entering the drinking population gravitate towards increasing consumption as a result of social pressures and mutual obligations.

Since the article by de Lint and Schmidt, several other

empirical studies of consumption levels in large populations have been used as evidence in favour of Ledermann's ideas. However, this research has also given rise to a statistical dispute regarding the validity of the specific predictions which can be derived from the Ledermann curve. For example, the underlying assumption of a fixed upper limit to possible consumption has been criticized and, thus, the important further assumption of a constant dispersion among log-normal distributions has been thrown in doubt. Criticism has also been made of the nature of some of the samples on which the Ledermann curve has been tested, including some of the original samples used by Ledermann himself. Some research on general population samples has simply failed to conform with predictions from the single-distribution model. Another kind of criticism has been to point to the weaknesses of working with national statistics, which may disguise important variations between sub-groups of the population. Certainly, in the individual case, the fixed relationship between consumption and the probability of alcohol-related damage does not hold up.

The consensus among specialists in this branch of epidemiology now appears to be that, although the single-parameter log-normal distribution is undoubtedly an over-simplification of the real state of affairs, it is a sufficiently good working approximation for present uses. Even if this is not accepted, no one could possibly dispute the broad, overall relationship between consumption and harm in a society. Statistical correlations between annual per capita consumption and rates of liver-cirrhosis mortality in a single country over time are some of the highest known to social science; the same correlations among different countries at the same point in time are equally striking. Correlations for other indices of problem drinking, such as hospital admission rates for alcoholism and drunkenness arrests, are less impressive but still high. What is more, these strong relationships apply in both directions—not only do cirrhosis mortality rates and other effects increase when per

capita consumption increases but, as was found during Prohibition in America and during the two world wars in many
European countries, they also drop when consumption falls.

This evidence is used mainly to argue for a policy of
prevention of drinking problems based on restricting the availability of alcohol through price increases or other measures.
We shall return to this crucial and highly topical issue (see
Chapters 5 and 9) but we are concerned here with the evidence's implications for the disease theory of alcoholism.

The first point is that all the work on the distribution of
consumption has shown it to be, whatever the exact shape of
the curve, 'unimodal' and not 'bimodal'; in other words, it has
one hump and not two. A bimodal distribution with a second
hump towards the right-hand end of the curves in Fig. 4.3
would be expected from a disease concept which postulated the
existence of a large group of moderate and relatively harmfree
drinkers and a small group of heavy-drinking and pathological
alcoholics. This is clearly contradicted by the evidence, which
shows that, at least in terms of quantities consumed, there is no
natural division between normal and abnormal drinkers and
that any distinction which is made for educational or treatment
purposes is arbitrary.

Admittedly, this is a somewhat simple-minded deduction
from the disease theory and may well apply only to fairly crude
disease concepts. There are presumably other ways in which
the drinking of alcoholics might be different from other drinking, apart from mere quantity. But however sophisticated the
disease theory appealed to, it must have difficulty with the
more general implication of the evidence on the distribution of
consumption. This is that the number of 'alcoholics' or problem
drinkers, however defined, varies closely with the average level
of consumption in a society. This is further evidence that
problem drinking is under environmental and not internally
located control. We now turn to a more direct comparison of
these two sorts of influence.

The influence of genetic and cultural factors

The evidence to be considered first in this section is perhaps the strongest to be found in favour of the disease theory but is still insufficient to prove it.[32] The idea that alcoholism is inherited has its roots in nineteenth-century thought and, like the disease concept of addiction itself, was rediscovered by Alcoholics Anonymous and the Alcoholism Movement in the 1930s and 1940s. That alcoholism was inherited fitted very conveniently with the AA assumption of a predisposition to contract the disease. After a period during the 1950s and 1960s when alcoholism researchers became preoccupied with socio-cultural explanations for differences in rates of problem drinking and the notion of an inherited predisposition fell largely into disrepute, it has recently become fashionable again owing to the sophisticated investigations of a number of Scandinavian researchers and the dedicated efforts of Donald L. Goodwin and his associates from the University of Kansas Medical Center.

The possibility that alcoholism may be inherited is suggested firstly by experiments with animals. The chief technique is that of 'selective breeding', in which a sample of rats is divided, for example, on the basis of their measured preference for alcohol. It has then proved possible to breed separate strains of rats among whom the difference in alcohol preference is gradually increased, thus showing that it is, to some extent at least, genetically determined. Differences in alcohol metabolism have also been selectively bred in this way. These studies raise the possibility that parallel characteristics may be inherited in humans, but the dangers of directly extrapolating from animal studies to conclusions about human behaviour are, of course, deservedly notorious. It is certainly true that alcoholism in humans tends to run in families. Alcoholic fathers tend to have alcoholic sons and problem drinkers in general tend to have problem-drinking relatives. The crucial question is why this happens—whether by genetic inheritance or through environmental example and influence.

The classic method of sorting out the relative contributions of genetic and environmental factors to a specified piece of behaviour is twin studies. Identical or monozygotic (MZ) twins, who have an identical genetic make-up, are compared with fraternal or dizygotic (DZ) twins, who are no more alike genetically than ordinary siblings. If the behaviour is genetically determined, one would expect it to show a greater degree of similarity, or 'concordance' as it is known, among MZ than DZ twins. In one Swedish study reported in 1960, this was indeed the case. When 174 male twin pairs were placed in five categories of drinking behaviour, the concordance rate among MZ twins was 53 per cent, while among DZ twins it was 30 per cent. When only those twins with a diagnosis of 'chronic alcoholism' were considered, the concordance rate increased to roughly 70 per cent. On the face of it, genetic transmission for alcoholism appears to be supported by this evidence.

Later research, however, has shown the situation to be somewhat more complicated than this. The largest twin study ever conducted was by Partanen and his colleagues from Finland, who reported in 1966 on 902 young male twins. Results showed that symptoms which were presumably related to alcoholism, such as drinking with loss of control, drunkenness arrests, and social complications due to excessive drinking, did not appear to be under genetic control. It was the frequency and regularity of drinking and the amount typically drunk at one session which showed a significant genetic influence. The highest genetic loading was for the dichotomy of whether subjects drank at all or were abstainers, a finding which has been supported in other research. Thus, accepting for the moment that there is some genetic causal influence on the likelihood of being diagnosed an alcoholic, it is not at all clear what precisely is being inherited.

Subsequent twin studies have produced varying results and no clear consensus of their interpretation has emerged. Even the basic finding of higher concordance rates for identical than fraternal twins has been challenged by preliminary findings from the Maudsley Hospital which give a concordance rate of 21 per cent for MZ and 25 per cent for DZ pairs. Moreover, there are considerable problems attached to the interpretation of data from twin studies, especially with regard to the assumption that it is only in degree of genetic similarity that MZ and DZ twins differ. For these reasons, alcoholism researchers have turned increasingly towards a different method of teasing out the influences of genetic and environmental factors. This is the study of the children of problem drinkers who have been separated from their parents shortly after birth and adopted. Clearly, if alcoholism is hereditary then children whose natural parents were alcoholics should show a greater than average tendency to become alcoholics themselves, irrespective of the drinking habits of their adoptive parents. On the other hand, if it is environmentally determined, the behaviour of the adoptive parents should be the main factor determining the behaviour of their children, irrespective of whether the natural parents were alcoholics or not.

This is the method that was employed by Goodwin and his colleagues, who also collected their evidence in Scandinavia.[33] In a first report, fifty-five adopted-away sons of alcoholics were compared with seventy-eight control adoptees whose parents showed no evidence of problem drinking. The main result was that the sons of alcoholics had a four times greater probability of having alcohol problems themselves than the sons of non-alcoholics. In a later report, the adopted-away sons of alcoholics were contrasted with their brothers who had been raised by the natural parents. Rates of alcoholism were very similar between these two groups, suggesting the greater importance of genetic than environmental factors. With the single exception of divorce rate, there were no other differences between sons of alcoholics and non-alcoholics, implying that what is inherited must be fairly specific to drinking behaviour and not some kind of generalized personality disorder which would lead to problem drinking in combination with a range of other disabilities.

The findings above are all concerned with the sons of alcoholics. The position with regard to daughters, and the role of hereditary mechanisms in female alcoholism in general, tells a very different story. Goodwin's data and those of other researchers have failed to reveal any differences between adopted-away daughters of alcoholics and control subjects or any other evidence suggestive of a role for the inheritance of alcoholism in women. This poses considerable difficulties for a genetic explanation. It is of course possible to propose that alcoholism is inherited by means of a 'sex-linked' gene (that is, attached to the Y but not the X chromosome) but this appears implausible. Another explanation put forward is that the same inherited, underlying disorder manifests itself as problem drinking in men but as depression in women, but this too is speculative. Some of Goodwin's data show that daughters of alcoholics reared by the natural parents are more prone to depression than those adopted by non-alcoholics, suggesting here the greater relevance of the kind of environment the girl was exposed to.

Apart from the puzzling findings on women, adoption studies are subject to other difficulties. It is often pointed out that selection processes may enter into the placement of adoptees, so that children of known alcoholics may have been placed in worse home environments than the control subjects they are being compared with. Thus, environmental conditions could be responsible for some of the differences observed. Furthermore, the possibility cannot be entirely ruled out that some adoptees come to learn of the drinking of their natural parent and that this affects their own behaviour in some way—perhaps in the manner of the self-fulfilling prophecy. But despite these reservations, it is probably fair to say that the accumulated evidence of twin and adoption studies, combined with evidence from animal work and some other kinds of research, has succeeded in demonstrating a role for inheritance in the genesis of problem drinking. The drinking of some kinds of problem drinkers is to some extent and in some way under genetic control. The contribution of genetic factors is best described as modest.

As to the precise way this contribution is made, it is very unlikely that alcoholism is transmitted by a single gene, whether it be dominant, recessive, or sex-linked. Far more likely, in view of the multi-faceted nature of the behaviour we call alcoholism or problem drinking and of the highly complex interactions of heredity and environment, is the view that many genetic influences are involved. In other words, inheritance is 'polygenic'. Despite Goodwin's data, it is still not possible to rule out the hypothesis that what is inherited is a generalized personality disorder or underlying psychiatric disturbance. The most frequently cited candidate is some kind of brain dysfunction which results in behaviour disorders among children and alcoholism and other forms of 'sociopathic' behaviour among adults. There is also, as we have seen, the suggestion that an underlying psychiatric disorder eventuates in problem drinking in men and depression in women. However, even the most enthusiastic advocate of these ideas would hardly claim that they applied to more than a proportion of those people who are labelled as alcoholics or problem drinkers.

Another group of hypotheses is centred on the assumption that it is differences in the way alcohol is metabolized which are under genetic control. One theory proposes that such differences result in some people having higher than average levels of acetaldehyde, a breakdown product of alcohol, in the blood. This in turn is thought to lead to a greater production of opiate-like substances which give the ingestion of alcohol a powerfully reinforcing effect for such individuals and thus predisposes them to excessive drinking. A different line of reasoning suggests that the liability to develop convulsions, delirium tremens, and other components of physical dependence on alcohol is genetically determined. Yet another possibility is that it is a differential susceptibility, given the same level of intake, to forms of alcohol-related damage which is inherited; and there is evidence that some problem drinkers are genetically predisposed to develop liver cirrhosis.

It is clear that there are several distinct ways in which heredity might make a contribution to the development of drinking problems. It is essential to recognize, however, that this is not a sufficient ground for calling alcoholism a disease. In the first place, to take the assumption of irreversibility, none of the 'pharmaco-genetic' mechanisms just described suggest that harmful drinking must inevitably get worse or can never be improved without total abstinence. Because a phenomenon is subject to hereditary influences does not mean it cannot be changed. For example, human weight has a strong genetic component but it does not follow from this that weight is unchangeable or that overweight people cannot be helped to reduce. Similarly, the concession that the development and mode of expression of alcohol problems are to some extent determined by hereditary factors in no way diminishes the relevance of a learning-theory account of drinking problems or the potential usefulness of methods of modifying drinking behaviour which may be derived from it.

In comparing the relative contributions of genetic and environmental factors, it has long been established that the two kinds of influence are not simply added together in their effects on behaviour. It is a total, one-to-one interaction which must be assumed. The only sensible question which can be asked is: How much of the variation in drinking behaviour is accounted for by environmental and how much by hereditary sources? What chiefly determines why one individual drinks frequently and heavily, while another drinks only occasionally and moderately, and yet another does not drink at all?

Put in this way, the answer to the question is clear. It is environmental factors which far outweigh the effects of possible genetic factors in determining differences in the way people drink. Research conducted mainly in the 1950s and 1960s demonstrated huge variations in rates of drinking and drinking problems among the various cultural and ethnic groups of America. A classic line of inquiry, associated chiefly with the sociologist, Robert Bales, compared Jews and those of Irish ancestry living in New York.[34] Although few were abstinent from alcohol the Jews had a very low rate of drinking problems. By comparison, the Irish had somewhat higher rates of abstinence but much higher rates of alcohol-related problems. It might be thought possible to explain these differences by appealing to genetic differences between races, but this is very far-fetched. Even a cursory examination of Jewish culture reveals how traditional rituals and ceremonial uses of alcohol serve to inculcate a moderate use and guard against the dangers of excess, whereas in Irish communities high positive value is often placed on drinking and intoxication. Many other examples of the cultural determinants of drinking could be provided to buttress this argument. Thus it is differences in the cultural fabric of social life which go far to deciding how often and how much we drink.

Apart from broad cultural influences, much later research had been devoted to showing the importance of variables like age, sex, educational level, and religious affiliation in determining levels of consumption and alcoholism rates. The kind of occupation followed is also of great importance. Thus, to repeat, even if it is accepted that genetic factors make some contribution to the development of drinking problems in some cases, they are greatly outweighed in importance by sociocultural and environmental factors in general. Finally, the great, world-wide increase in alcohol problems, which has been called 'the alcoholism epidemic' (p. 8), is far more likely to be due to economic and socio-cultural changes over time than to any changes in the gene pool.

Conclusions

Let us now attempt to summarize the implications of the evidence covered in this chapter for the disease perspective on problem drinking. One way of doing this is to reconsider the 'syndrome' of disease-theory assumptions listed at the end of Chapter 3 and decide how the evidence affects an evaluation of each.

1. A great deal of evidence has been produced—from studies of resumed normal drinking, laboratory investigations of drinking behaviour, household surveys, and data on the distribution of alcohol consumption—to show that it is impossible to draw a hard and fast line between 'alcoholics' and 'non-alcoholics'. The drinking of people labelled as alcoholics or problem drinkers is subject to the same kind of laws and changes in response to the same kind of influences as the drinking of people not so labelled. Thus problem drinking is not a discrete entity but is continuous with apparently normal or non-problem drinking.

The corollary of this is that there is no single constellation of alcohol-related problems which could be described as 'alcoholism'. Rather, there is a range of problems—medical, legal, occupational, social, marital, or interpersonal—which can be experienced by those who drink excessively and the particular problems experienced will depend on the particular circumstances of the individual case.

2. The traditional concepts of loss of control and craving are unhelpful in the explanation of harmful drinking, as shown

especially by studies of the actual drinking behaviour of diagnosed chronic alcoholics observed under controlled conditions. Alcoholic drinking, like any other goal-directed activity, is 'operant' behaviour and can be altered by changing its environmental consequences. Even as a description of alcoholic drinking, 'loss of control' is in need of considerable modification. It is again impossible to distinguish sharply between alcoholic and non-alcoholic drinking by pointing to a sense in which the latter is completely within voluntary control and the former outside it.

3. Although versions of the concepts of impaired control and craving may have some validity in the description of the behaviour and experiences of some problem drinkers, there is no evidence that such processes are irreversible. Studies of resumed normal drinking after treatment and household surveys of drinking in the natural environment have demonstrated that drinking problems in general are not irreversible phenomena. Present knowledge suggests that many serious-problem drinkers should be strongly advised to abstain totally from alcohol, but controlled drinking is a viable treatment goal for other problem drinkers, especially those with less serious problems.

4. As implied by the potential reversibility of problem drinking, there is no evidence of an inevitable, progressive deterioration in the problem drinker's condition if drinking is continued. Earlier evidence of a relatively inexorable sequence of 'symptoms' or phases of alcoholism has not been borne out by the evidence. While treatment may be of benefit in individual cases, its effects on the lives of many problem drinkers are insignificant compared with the influence of life-events and environmental changes.

The evidence in this chapter may also be partly summarized by referring to the three types of disease concept described in

Chapter 3. To the extent that these disease-theory types share
the assumptions which have just been discussed, they are
subject to the same criticisms as the disease perspective in
general. However, by considering each type in turn, some
additional points may be raised for discussion.

Type A: Alcoholism as a pre-existent physical abnormality.
Despite the fact that this is the type of disease theory which is
most familiar to the general public, the evidence shows that it
is by far the weakest. The only research findings which could
conceivably be advanced to support it are the twin and adoption
studies which suggest the possibility of a limited role for genetic
factors in the causation of drinking problems. However, even if
some kind of genetically determined, pre-existent abnormality
is assumed to exist among a minority of diagnosed alcoholics,
little progress has been made in identifying it.

Against this, a whole range of evidence has been produced
to invalidate this type of disease theory, including all the
research which demonstrates that, in one way or another, a
strict separation between the drinking of alcoholics and non-
alcoholics is simply not justified and that drinking problems are
not irreversible in principle.

Type B: Alcoholism as mental illness or psychopathology. Apart
from the evidence covered in this chapter, theories in this
category have difficulty with the inconsistencies and logical
objections mentioned in Chapter 3 (see pp. 74–8). Problem
drinkers have not been found to show any particular kind of
psychopathology in their early learning experiences or to demonstrate any special personality type which renders them
uniquely susceptible to alcohol problems.

It must be recognized, however, that theories of this type are
much more flexible than those considered under Type A. They
may well be based on some notion of a continuum between
normal and abnormal functioning rather than a qualitative

distinction between alcoholic and non-alcoholic. Some psychoanalysts and psychotherapists have explicitly stated that drinking problems are not necessarily irreversible and that normal drinking can sometimes be taken up after the successful resolution of the underlying neurotic disturbance. It would appear then that this liberal type of mental-illness concept has not been completely invalidated by the evidence.

The question which arises here, however, is that, if abnormal drinking is not regarded as qualitatively different or irreversible, what is the point in calling it a disease? Whether one calls something a disease or not is not just a semantic issue; it has profound consequences for how the alleged victims of that disease make sense of their own behaviour and for how society responds to their problems. The use of the word 'illness' in the concept of 'mental illness' is a metaphor which has been transferred from the organic realm of physical disease to the area of experience and behaviour. By the concept of 'mental illness', we are invited to think of a piece of behaviour as though it were the same kind of thing as a physical illness. The question then becomes: What are the consequences for the person concerned and for society in general of using this metaphor in this particular context? In other words, it becomes a question of the advantages and disadvantages of employing a disease concept to describe abnormal drinking behaviour. This is the subject of the following chapter.

Type C: Alcoholism as acquired addiction or dependence. The same kind of considerations apply to the last type of disease theory, the one which is currently most respectable among academics and researchers. In so far as theories in this category subscribe to the four assumptions discussed above, they will clearly be embarrassed by the evidence reviewed in this chapter. Again, however, more sophisticated theories in this class do not necessarily assume discontinuity or irreversibility. If alcohol dependence is equated with physical addiction to alcohol in the same way as the drug addict is addicted to heroin, this appears to form the basis for a disease concept which avoids the major implications of the evidence reviewed.

Once more, it all boils down to a question of whether it is useful to use the term 'disease'. There are two reasons to suspect that it might not be. First, although addiction or physical dependence on alcohol is undoubtedly a factor in the maintenance of a drinking problem, there are grounds for believing that its importance has been exaggerated in the traditional perspective. 'Detoxification', or the removal of physical dependence, is now a relatively simple affair and, if that were all there was to solving a drinking problem, one could guarantee almost 100 per cent success. The real difficulties, of course, only start after detoxification when the problem drinker returns to his or her home environment and the persons and situations associated with excessive drinking in the past. It is the prevention of relapse, or a return to harmful drinking, which represents the main task for treatment or the problem drinker's efforts at self-help. The evidence suggests that relapses occur, not directly because of physical dependence, but because of social and psychological factors in the form of life events, lack of coping skills, and conditioned responses to alcohol-related cues—in short, because of aspects of problem drinking which have been learned. Thus, the abolition of physical dependence is a relatively unimportant part of the solution to a drinking problem and it is not helpful to exaggerate its importance by making it the corner-stone of a disease concept.

Even if this were not the case, and the predominance of physical dependence in explaining and treating alcoholism were still insisted on, it nevertheless would not follow that it was useful to call problem drinking a disease. Consider the example of smoking. Here is a piece of behaviour which undoubtedly causes a huge amount of personal and social damage and which is also undoubtedly perpetuated by physical dependence on nicotine, probably more than alcoholism is maintained by physical dependence on alcohol. But nobody seriously suggests thinking of smoking as a disease, presumably because enormous financial and legal complications would arise. In other words, there would be considerable disadvantages from doing so. The position with respect to dependence on alcohol is exactly the same. Calling alcohol dependence a disease is a decision which must be defended in terms of the advantages and disadvantages of doing so. This leads conveniently to the subject matter of the next chapter.

5 Pros and cons of the disease perspective

In his classic discussion of the disease theory of alcoholism,[1] E. M. Jellinek made the following statement: 'It comes to this, that a disease is what the medical profession recognizes as such.' With the benefit of hindsight, we may now conclude that Jellinek was both right and wrong when he made this much-quoted remark.

Jellinek was right in the implication that whether or not something should be termed a disease is an essentially practical decision, based on the consequences of doing so. Thus, the proper question for consideration is, not whether alcoholism (or problem drinking) is or is not a disease, but whether it is practically useful to *regard* it as a disease. Given that applications of the concept have varied widely in different places and times, a precise definition of 'disease' is an extremely complex matter which remains to be resolved; in the meantime we are faced with the pressing task of responding to an expanding number of drinking problems in our society. Hence the importance of debating the various pros and cons which flow from declaring problem drinkers to be sick persons.

Where Jellinek was wrong was in his presumption that it was only the medical profession which should have a say in making this decision. Since he wrote, the phenomenon known as 'medical imperialism'—the tendency of sections of the medical profession to claim increasingly diverse types of human problem as their own special areas of expertise—has become well recognized. This is especially true with regard to deviant behaviours, or ways of behaving which powerful groups in society find unwelcome and wish to eradicate or control. In the areas of criminal and sexual deviance, the medicalization of such deviant behaviours has been a feature of industrialized societies for the last 200 years or so. More recently, however, the tide of medical imperialism has been stemmed to some extent by the realization that medical claims to expertise are sometimes spurious and that, in any event, the whole of society is affected by the medicalization of deviance and should therefore be consulted in any decision to apply the label of disease to an area of human behaviour.

Attitudes to the disease concept of alcoholism

The main advantage claimed for the disease view of problem drinking is the most general one, that it has improved the lot of the problem drinker. We have seen that, despite its being presented with a veneer of scientific respectability, the chief intention behind the new disease approach to alcoholism in the 1930s and 1940s was to obtain a better deal for the suffering alcoholic, in terms of very practical matters like legitimate access to treatment services and insurance benefits, and avoidance of imprisonment and other harsh consequences of public drunkenness. Less tangible benefits such as a more sympathetic attitude on the part of relatives, friends, and employers and a reduction in the stigma associated with chronic drunkenness were sought too. In terms of formal criteria like official recognition by medical bodies and other authorities, the Alcoholism Movement must be rated a definite success. In addition, there can be no doubt that the effort to have alcoholism recognized as a disease has succeeded in keeping many alcoholics out of prison and has indisputably added to their welfare compared with the situation before this effort began. But all this having been conceded, how successful has the Alcoholism Movement been in engendering sympathetic attitudes among important professional groups and among the public at large?

How successful has it been in combatting stigma? How far has it succeeded, in other words, in its avowed aim of replacing a moral conception of alcohol problems with a genuine disease conception?

There is a body of evidence which is relevant to this question, consisting largely of the results of surveys and opinion polls of the general public and more specialized groups. One immediate conclusion from this evidence is that the proportion of people found to agree with the bald statement that alcoholism is an illness or that the alcoholic is a sick person steadily increased after the Second World War. At the end of the 1940s, roughly one fifth of members of the general public interviewed agreed with such statements, whereas in the early 1960s the proportion had risen to about three fifths.[2] It is not clear whether it has continued rising or whether a ceiling has been reached, but more people are now prepared to accept the statement that alcoholism is a disease than before the Alcoholism Movement began its propaganda campaign.

But is this really a straightforward triumph for the Alcoholism Movement's efforts at public education? When the evidence is examined more closely, there emerge several reasons for believing it may not be. First, one study[3] found that those who endorsed the disease concept were likely to score highly on a scale of 'social desirability'; that is, they were people who were anxious to create a good impression. The possibility arises therefore that these people were merely paying lip service to the notion that alcoholism is a disease. This possibility is strengthened by an inspection of what endorsement actually means for most people. Acceptance of the idea was supposed to carry with it the implication that the alcoholic could not be held responsible for his behaviour; it would therefore be expected that, as the disease concept gained in acceptance, the number of those believing that alcoholism was a form of moral weakness would decrease. But this is not at all what has been found. For example, in a survey conducted in Iowa in 1961,[4] the proportion of those defining the alcoholic as a sick person had reached 65 per cent, but 75 per cent of the same sample defined him also as 'morally weak' or 'weak-willed'. That many people find it possible to hold both, seemingly incompatible views is shown by the fact that no less than 41 per cent of the entire sample defined the alcoholic as both morally weak and sick! Thus there is good evidence that the growth in public acceptance of the disease concept has not been accompanied by a corresponding decline in the moral weakness view of alcoholism and that the major intended effect of the Alcoholism Movement's campaign of education has not worked. There is even evidence to suggest that those who subscribe to a mental-illness view of alcoholism are more likely, rather than less likely, to agree that it is a form of moral weakness—a curious finding perhaps explained by the fact that, in common parlance, calling someone mentally ill is one of the most insulting things one can say. However this may be, it seems clear that, for a substantial proportion of the population, the disease perspective has not succeeded in absolving the problem drinker from moral responsibility for his or her actions.

A recent study by John Crawford and one of the present authors[5] throws more light on this issue. In a household survey of 200 members of the general public, a modest but statistically significant positive relationship was found between believing that alcoholism was a disease and agreeing that alcoholics deserve sympathy and help; those who believed the former were more likely to accept the latter. However, on closer inspection of the data, it appeared that this relationship was simply a spurious consequence of the fact that both the above attitudes were independently related to a third factor, a rejection of the idea that alcoholism was a sign of moral weakness. Thus, those who believed in the disease concept were less likely to adopt the moral-weakness point of view which led them, in turn, to hold a more sympathetic attitude to alcoholics. When the effects of the moral-weakness factor were extracted from

the statistical analysis, the relationship between the disease
concept and a more sympathetic attitude disappeared. These
results show that whether or not one believes in the disease
concept of alcoholism has nothing directly to do with whether
or not one is likely to adopt a sympathetic attitude to the
treatment of alcoholics. Rather, it is likely that this attitude is
associated with the trend during the twentieth century towards
non-condemning, humanitarian attitudes to deviant groups in
general.[5]

The situation with regard to the attitudes of professional
groups involved in the response to problem drinking is hardly
more encouraging for the disease perspective. Although one
study published in 1970 found 85 per cent agreement among a
group of social workers with the statement that alcoholism is a
disease, another in 1971 discovered that only 35 per cent of 925
psychiatrists and psychologists endorsed this position.[2] Based
on open-ended interviews, a study in Scotland of the attitudes
of general practitioners,[6] on whom so much of the initial
response to problem drinking depends, revealed that the majority did not hold clearly medical conceptions of alcoholism but
tended to possess ambiguous and ambivalent attitudes centred
around notions of 'will-power' and social and environmental
pressures. Thus, even among professional groups whose attitudes are of great significance for the provision of humane
treatment services for problem drinkers, the disease perspective
has proved far from an unmitigated success.

We may conclude that the Alcoholism Movement has had
only limited success in its attempt to propagate a disease view
of alcoholism in our society and has achieved few of its original
aims. This great effort has met with much formal recognition
and has undoubtedly led to some practical improvements in the
alcoholic's position, particularly with respect to a diversion
from a penal to a treatment response among 'skid-row' alcoholics. However, there is little sign that the main target of
replacing a moral with a disease view—one which amounts to

any more than fashionable lip-service to repetitive slogans—
has met with much success. Meanwhile, the disease theory
continues to have concrete disadvantages.

Disadvantages in treatment

There is a certain irony in the use of the word 'treatment' in
the title of this section, since this itself implies some kind of
medical procedure. The fact is, though, that other words or
phrases are too long-winded. All that is meant by treatment
here is the effort, on the part of psychiatrists, social workers,
psychologists, or any other group of helpers, using any kind of
helping techniques, to intervene benignly in the lives of problem drinkers. The argument of this section is that the disease
perspective can be seen to impede this effort in several ways.

Responsibility for change

Perhaps the chief connotation of the term 'disease' is that it
describes things that happen to us, not things over which we
have, or could have, control. Diseases or illnesses are things
we 'catch' or acquire in some other way; they are seen as
external to ourselves and as having nothing to do with us
personally. Since we are not generally responsible for their
arrival, we do not usually hold ourselves responsible for their
subsequent course. By the same token, we assume there is little
we can do about illnesses apart from co-operating with medical
treatment. When we become ill, we hand over that part of the
responsibility for ourselves which is connected with our illness
to the medical expert, on the understanding that this is the best
way of becoming cured. The very word 'patient' implies that
we must wait passively for the medical intervention to do its
work and return us to a normal, healthy state.

All these attributes of disease may be perfectly necessary in other contexts, but in the attempt to solve or ameliorate drinking problems, they are distinctly unhelpful. Even the proponents of the disease theory recognize that the patient must be an active participant in the change process, despite the fact that the connotations of their own theory all contribute to passivity. Indeed, there is good evidence that problem drinkers treated with tranquillizers do less well than those whose treatment does not involve any obviously medical ingredients.[7] Successful treatment for alcohol problems must be based on a partnership between therapist and client in which the latter gradually takes over responsibility for control of his or her drinking. But so strong are the expectations derived from the vocabulary of disease that often the first task for the therapist is to disabuse clients of the idea that treatment will involve something being 'done' to them and that there is some kind of magical cure for their problems.

What is true of the initiation of behaviour change also applies to its maintenance over time—the client must be persuaded to take responsibility for the retention of any gains which have been made. A well-established finding from social psychology is that behaviour changes which are credited to oneself are maintained much better than those attributed to some external force or agency, such as the effectiveness of treatment received. In accordance with this principle, the aim of treatment should be to increase clients' feelings of 'self-efficacy' (see p. 258) in relation to their drinking and other life problems. So far from this being incorporated in the disease theory, the irony is that when the client improves it is usually regarded as a success for treatment but when he or she does not, it is because of a failure to comply with the treatment regime, poor motivation, or some other inadequacy of the problem drinker him- or herself. As William Miller has pointed out in a thoughtful discussion of treatment for problem drinking,[8] in the disease model all questions of motivation are seen as individual properties of the client's personality rather than being placed in the context of a negotiation between helper and helped.

Apart from the more general connotations of disease, disadvantages also accrue from the more specific content of the disease theory of alcoholism. The most prominent is concerned with the central notion of 'loss of control' and the associated maxim of 'one drink, one drunk'. This encourages problem drinkers to abandon personal responsibility for their drinking by teaching them that, owing to the nature of the disease they are unfortunate enough to be suffering from, if they ever take a drink while attempting to abstain, they will be unable to prevent themselves from getting completely drunk. It is quite natural that when treated problem drinkers experience a slip, as so many do for a variety of reasons, they will conclude that they 'may as well be hanged for a sheep as a lamb'. Any guilt they may be prone to experience as a result of this lapse can be effectively assuaged by an appeal to the disease concept. This psychological process has been expressed in more formal language by Alan Marlatt in his concept of 'the abstinence violation effect' (see pp. 264–5). As already pointed out, it is difficult to overestimate the force of expectations and beliefs about alcoholism in affecting the problem drinker's behaviour. In research by one of the authors and his colleagues,[9] it was found that, among a group of problem drinkers who had taken a drink during the six months following abstinence-oriented treatment, it was beliefs about alcoholism, rather than the estimated degree of dependence on alcohol previously reached, which determined whether they would go on to drink to excess and relapse or would manage to maintain their drinking within harm-free limits.

Labelling

The act of calling a piece of behaviour a disease does much more than merely give it a name; it actually changes the behaviour in question. This broad insight has been formally stated and elaborated in a branch of sociological theory, which became popular during the 1960s, usually known as *labelling theory*.

A fundamental distinction in labelling theory is between primary and secondary deviance. *Primary deviance* refers to the original breaking of social rules before society has had an opportunity to react. It is assumed to occur on a fairly random basis and to be the result of a great number of different causes. *Secondary deviance* refers to any subsequent rule-breaking which takes place as a consequence of changes in the deviant person's identity caused by societal reaction to the primary deviance. With respect to alcoholism, it is assumed that many people drink in ways which other people would find offensive and would define as abnormal. However, only some of this primary deviance becomes labelled as alcoholism and a target for corrective intervention.

Having been persuaded to accept a self-definition as an alcoholic, the person then undergoes changes in identity which may aggravate and perpetuate the deviant drinking. First, the diagnosed alcoholic may take on 'the sick role' in which, in return for an exemption from normal social obligations, he is expected to co-operate passively with whatever medical procedures are directed at him. The main advantage to the alcoholic individual from this is that he is not held responsible for his deviant actions; at the same time, however, excessive drinking is legitimized and reinforced. Second, expectations arising from the disease theory further legitimize deviant drinking and provide a ready explanation for failure to change. Third, through his experiences of the treatment regimes to which he is exposed, the diagnosed alcoholic may come to associate more and more with other alcoholics, who are more tolerant of his drinking behaviour. This leads to further deleterious changes in self-concept and self-esteem. Finally, the individual may come to adopt a 'career' as an alcoholic patient which may last for the rest of his life.

It might be objected that the problem-drinking paradigm being proposed as an alternative to the disease model will result in the same labelling tendencies. There is some truth in this, in so far as any attempt to understand and analyse problem drinking will inevitably lead to a categorization of behaviour. However, there are important differences between the two perspectives which lessen the labelling propensities of the learning-theory approach. In the first place, there is far less emphasis on getting the person finally to admit to being an 'alcoholic', with the great battle to overcome the patient's 'resistance' and 'denial' this entails. This attempt to force an acceptance of the alcoholic label becomes necessary only if the extreme solution of total and lifelong abstinence is being insisted on and, even then, it is not at all clear that it is absolutely essential to progress. Second, many of the methods of intervention which derive from the learning-theory approach manage to avoid the ponderous labelling processes of formal medical diagnosis and treatment. The much greater emphasis on prevention through public-health policy avoids the issue of labelling entirely and this is one of its main advantages.

Perhaps the most damaging consequence of labelling is the stigma it gives rise to. It is most ironic that the disease model, which claims to engender more sympathetic attitudes towards the problem drinker, often results in the very opposite, the tendency to regard diagnosed alcoholics as undesirable and possibly dangerous outcasts from normal society. This is especially likely to happen when Alcoholism Treatment Units are located in psychiatric hospitals, with all the fearsome associations such institutions possess for many people. This tendency is inevitable given that the deepest purpose of the labelling process is to provide us with scapegoats on which to blame the ills of society. As the sociologist Dan Beauchamp has pointed out,[10] it is remarkable how willing AA affiliates seem to be to accept the role of outcast and, indeed, compound it by their obsession with anonymity and secrecy. AA offers the potential recruit a unique status as a 'recovering alcoholic' and a feeling of being set apart from the rest of the human race. The obverse of this coin is that the AA affiliate must be prepared to acknowledge the stigma of a permanent handicap in living.

The exclusive demand for abstinence

A whole set of disadvantages are attached to the disease theory because of its insistence that abstinence is the only solution to a drinking problem. As already made clear, there are circumstances where total abstinence is necessary for recovery. There are many other circumstances when it is not and, by ignoring these, the disease theory unnecessarily restricts the overall treatment response to problem drinking.

The chief way it does so is by restricting attention to the upper end of the spectrum of seriousness or dependence where abstinence may indeed be necessary. But by promoting the image of alcoholism as involving only the most dramatic manifestations of physical dependence, such as delirium tremens, hallucinations, and convulsions, and the most serious social consequences, such as complete disintegration of career and marriage, the impression is given that this is the only form of problem drinking worth considering and that anything less is too trivial to be concerned about. What this ignores, of course, are all the many kinds of alcohol-related damage which are experienced by large numbers of people in our society and which, precisely because they do not involve severe dependence, are more amenable to change. This is not to deny that the problems of the severely dependent deserve anything less than our full compassion and help. (Indeed, we would go further and argue that *more* resources should be provided for chronic and severely disabled problem drinkers.) At the same time, this image of alcoholism should not be allowed to overshadow the many other forms of alcohol-related damage which occur.

The main reason for the failure of the disease theory to come to grips with the full range of problems is clearly because of its exclusive reliance on abstinence. The AA concept of 'rock-bottom' is based on the recognition that a problem drinker must experience a great deal of suffering and reach a very low ebb before being able to contemplate lifelong abstention. In fairness, it should be admitted that therapists working in the disease tradition do attempt to persuade people to accept treatment before the greatest degree of harm has been done. But this almost invariably entails a confrontational posture by the therapist and the exhausting battle with the patient to extract a surrender to the diagnosis of alcoholism mentioned in the previous section. This battle is necessary because, in our present-day society, asking someone to become a total and lifelong abstainer is asking a great deal.

This is especially true of young people and it is worth remembering that the highest incidence of alcohol problems, as shown by the evidence of household surveys (see p. 111), is among men in their 20s. In contrast to this, the mean age of patients in specialist alcoholism-treatment services is usually over 40 and it is also relevant that AA has not succeeded in its aim of attracting younger people or those earlier in the course of their problem (see p. 55). This is the most obvious drawback of the disease theory in treatment; by ruling out the possibility of some form of continued drinking as an explicit goal, it cannot attract problem drinkers into treatment before serious levels of damage have occurred. Thus, it is in the area of early intervention and, to use the medical term, 'secondary prevention' (the prevention of more serious problems from occurring among those already affected) that controlled-drinking treatment techniques come into their own.

We also know from household survey data that only a proportion of those with mild or moderate problems go on to develop more serious difficulties with alcohol. Even accepting, though, that the rate of spontaneous recovery is high, such lesser problems do impair people's lives and should be regarded as legitimate targets for intervention, especially since the evidence suggests that there is a good chance of being able to modify them. A survey conducted in Scotland in the early 1970s[11] showed that 18 per cent of men interviewed expressed a desire to drink less, giving some indication of the huge numbers of individuals who might respond to some form of controlled-drinking intervention. The trouble is that there are so few services aimed at this target population and this is because the dominance of the disease theory has prevented controlled drinking from being acknowledged as a legitimate treatment objective. There is no need for this objective to be confined to traditional methods of treatment delivery, such as out-patient clinics, group therapy and the like. It can be located more profitably among community resources or rely on fairly simple self-help procedures delivered by various means (see pp. 296–303).

Even when younger or less serious problem drinkers are attracted into conventional treatment, there is good evidence that an abstinence goal yields an unnecessarily high rate of failure and is therefore inappropriate. The data from the Rand four-year follow-up (see pp. 94–6) showed that young, unmarried men with a low level of dependence were ten times more likely to relapse if they had adopted abstinence as a goal than if they had become non-problem drinkers eighteen months after treatment. A study by Martha Sanchez-Craig and her colleagues of the Addiction Research Foundation of Toronto[12] demonstrated that non-serious problem drinkers were more likely to keep to a rule of abstinence and to avoid heavy drinking during the treatment period if their ultimate goal was controlled drinking than if it was total abstinence. The authors considered controlled drinking to be the more appropriate goal for this type of problem drinker because it was more acceptable to the majority of them and most of those assigned to abstinence had in fact developed moderate drinking patterns on their own. Moreover, among the heavier drinkers taking part in the study, imposition of the goal of abstinence was ineffective in promoting abstinence and counter-productive in encouraging moderate drinking.

The reasons why an abstinence goal is counter-productive in these circumstances are not difficult to find. In the nineteenth century, when the demand for total abstinence from alcohol first flourished, it formed part of a familiar way of life. Large numbers of people were abstinent and they included some of the leading members of society. The recovered drunkard could easily find shining examples of sobriety to emulate and could expect considerable encouragement and praise for his new resolve. Moreover, abstinence was buttressed by a prominent set of social values centred round the virtues of hard work, self-discipline, and sturdy independence, with success in life assumed to be a sign of these moral qualities. Thus the reformed drunkard had a ready-made identity with which to equip his new life.

In our own century, things have changed dramatically, in a way which is bound up with the decline of organized religion. The work ethic and rigid morality have been largely replaced by a preoccupation with consumption, leisure, and interpersonal relationships. In consequence, the evidence clearly shows that the numbers of teetotallers in Western industrialized countries has been steadily decreasing since the end of the Second World War. There is also evidence that older attitudes of disapproval of drinking and even of occasional drunkenness have waned considerably[11] and that this has been accompanied by a greater degree of suspicion and even outright hostility towards those who choose not to drink.

For all these reasons, the position of the abstinent individual in our society is increasingly that of the social outcast and this very fact creates special difficulties for those to whom abstinence is a solution to a previous drinking problem. Several studies have shown that the successful maintenance of abstinence is not always accompanied by corresponding improvements in the overall quality of life. In one famous study reported in 1962,[13] abstinent alcoholics who could be termed 'independent successes' were in a small minority, the remainder being either overtly disturbed, inadequate, or entirely reliant on AA. This is not to claim that a controlled-drinking outcome is always associated with a better quality of life; the relationship between drinking and general life-adjustment is complicated. One thing is sure, however: total abstinence does not always lead to fulfilment in all aspects of life.

This brings up a further unfortunate consequence of an exclusive reliance on abstinence. From the traditional point of view, if the patient continues to drink excessively after treatment, he or she is usually regarded as a failure or is put somewhat grudgingly into the category of 'drinking but improved'. As psychiatrist Mansell Pattison had argued over a number of years,[14] this obscures the fact that, in cases where it is unrealistic to ever expect total abstinence or even harm-free drinking, a more modest reduction in drinking, if accompanied by improvements in social, vocational, or health adjustment, is a worthwhile and valuable treatment objective in its own right. Pattison has coined the term 'attenuated drinking' for an explicit goal of treatment in those, mainly skid-row alcoholics who have such major life problems that radical changes in drinking behaviour are virtually unobtainable. 'Harm reduction' is becoming an increasingly important aspect of the treatment agenda.

As well as being the goal of treatment, abstinence is also seen from the disease perspective as an essential precondition for treatment to take place and this too has been the subject of criticism. It has been pointed out that, if alcoholism really is a disease, it is absurd to attempt to treat it without allowing its major symptom, drinking, to take place. In learning-theory terms, the point is that a piece of behaviour cannot be extinguished or modified if it is not allowed to happen. It is well known that few alcoholics drink during their stay in hospital, despite typically having ample opportunities to do so. Yet a large proportion return to drinking within a short period after discharge. If asked, nearly all hospitalized alcoholics will deny experiencing craving and will be highly confident of having been 'cured'. As a result, few will have made plans for dealing with the craving which will suddenly re-emerge as soon as they return to their natural environment and encounter cues associated with drinking in the past. Although there will be considerable practical difficulties involved at first, it can reasonably be argued that drinking behaviour itself should be the main focus of concern in treatment and must not only be allowed to occur but positively encouraged, so that associated emotional responses and motivations for drinking can be understood and a new pattern of behaviour developed. All this is quite compatible with the learning-theory approach to treatment but completely antithetical to the highly artificial kind of treatment favoured in the Alcoholism Treatment Units inspired by the disease theory.

The claim is that a socio-psychological, learning paradigm of problem drinking leads logically to a variety of highly practical treatment procedures which may be tailored to the specific requirements of the individual case. By contrast, treatment based on some unspecified disease process has few implications except that alcoholics must never again touch alcohol. In other words, it cannot tell us what we should actually *do* with problem drinkers. No wonder that disease-oriented treatment for alcoholism has been found to consist of little more than commonsense cajoling and persuasion.[15] The disease theory has little else to offer.

Legal confusions

It will be recalled that the main political objective of the rediscovered disease concept put forward by the Alcoholism Movement was to improve the position of alcoholics by persuading society that they were not responsible for their drunken behaviour and therefore should not be punished for it. To the extent that the movement has been successful in this, it would be expected that the disease concept would be increasingly accepted by the courts and increasingly employed as a defence against various wrongdoings. Indeed, during the 1960s the disease concept was debated at length in the United States Supreme Court but, rather than clarifying the legal implications of the disease concept, this debate has given rise to considerable confusion.[16]

Legal discussion of the meaning of disease in the alcoholism context was preceded by an important case in the area of opiate addiction. In *Robinson* v. *California* in 1962, the Supreme Court found that drug addiction was indeed a disease and that this disease entailed a loss of control over use of the drug. Hence, any penalties which might be inflicted on the drug addict for displaying a symptom of a disease, that is, for drug-taking, amounted to 'cruel and unusual punishment' and were unconstitutional under the Eighth Amendment. Applying the same logic to the alcohol field, two cases heard in 1966 were both decided in favour of men who had been diagnosed as alcoholics. In the *Easter* and in the *Driver* cases, the Supreme Court reasoned that, since alcoholism had been defined as a disease by medical authorities and since a symptom of this disease was repeated public drunkenness, the defendant could not be punished for such drunkenness. In the last-mentioned case, the Court emphasized that this ruling in no way absolved alcoholics from the consequences of other illegal acts committed while intoxicated; exemption from punishment applied only to the drunkenness itself.

As well as ruling punishment for drunkenness unconstitutional, the decision in the *Driver* case addressed the question of what should be done with persons suffering from the disease of alcoholism. It recommended some form of appropriate detention for alcoholics for the purposes of treatment and rehabilitation and this recommendation was shortly afterwards affirmed in the *Easter* case. Unfortunately, the immediate effect was to cause chaos in the streets of Washington, DC, where the *Easter* ruling applied, because drunks were diverted from gaol without any adequate treatment facilities being available. The practical consequence was that homeless alcoholics were simply set free without experiencing either imprisonment or treatment.

Dissatisfaction with this state of affairs culminated in a legal wrangle in the case of *Powell* v. *Texas*, heard in the Supreme Court in 1968. On this occasion, the Court overturned the earlier decisions by finding in a majority of five to four against the defendant. The grounds were, first, the practical difficulties of implementing the previous decisions and the fear that to do so would result in 'thousands of alcoholics . . . roaming the streets'. Second, the justices cited the confusion and lack of agreement among medical authorities themselves over the precise meaning of the statement that alcoholism was a disease. Interestingly, they took special note of those logical difficulties applying to the concept of 'loss of control' discussed in Chapter 3. Finally, the Court was concerned about the undesirable implications of the previous decisions for the entire doctrine of criminal responsibility.

Although the decision in the *Powell* case was clearly a major rebuff for the disease theory, its protagonists could draw some comfort from the minority dissenting opinion of four of the justices. However, even this was made the occasion for a closely argued critique of the disease concept by the philosopher, Herbert Fingarette, in an article entitled 'The Perils of *Powell*'.[17] Fingarette was worried that the concepts of 'involuntariness' and 'disease' were slippery slopes on which to build a legal doctrine of withholding punishment from alcoholics, considering the logical inconsistences and lack of empirical support for the disease theory. He did not argue against offering humane and rational treatment to those suffering from alcohol problems but only that the purportedly medical knowledge of the disease of alcoholism was inadequate grounds for doing so.

Disadvantages for primary prevention

Probably the most outstanding disadvantages of the disease theory arise in the area of the prevention of drinking problems. Understandably enough, the main effects of the disease perspective have been to persuade the medical profession and the policy-makers that the correct response to the growing tide of problems is to provide more and better treatment facilities, and this belief inspired the enormous injection of money into alcoholism-treatment services in the United States during the 1970s. To the extent that there were any implications for prevention rather than treatment, they lay in the hope of early detection of those vulnerable individuals liable to develop the disease of alcoholism and a massive research effort aimed at discovering the crucial biochemical abnormality believed to be the cause of the disease. By any system of accounting, this method of responding to the 'alcoholism epidemic' has completely failed to arrest it and must be judged a dismal failure.

For a number of years now, there has been a strong and growing body of opinion among alcoholism specialists that this set of disease-theory assumptions are not only ineffective but dangerously misleading. These experts point to the indisputable evidence of very high levels of correlation between the per capita consumption of a population and the extent of various kinds of alcohol-related damage. As noted in Chapter 4, these correlations are among the highest known to social science and are equally impressive in statistics collected over time in the same country as over a range of countries at the same point in time. Although all the mathematical assumptions and deductions associated with the Ledermann formula (see p. 128) may not be exactly right, it is nevertheless true that, if the average consumption of a given population is known, the number of people drinking over a certain level—say, the level at which the risk of liver disease begins to rise or the level at which physical dependence on alcohol begins to develop—can be calculated with reasonable accuracy. Moreover, a given increase in average consumption leads to a disproportionately large increase in the number of excessive drinkers. For all practical purposes, there is a fixed relationship between the average alcohol consumption found in a society and the amount of damage caused by alcohol in that society.

It follows from this well-established evidence that the most direct method of reducing alcohol-related damage is simply by reducing average consumption. How can this be done? Here, supporters of this point of view refer to a further body of evidence relating per capita consumption to the availability of alcohol in general and its retail price in particular. There is no doubt that the large increase in consumption and problems found in many countries of the world over the last thirty or forty years has been associated with a substantial decrease in the real price of alcoholic drinks. There is also evidence that, when alcohol becomes less available or more expensive, consumption and problems decline. So, put at its very simplest, the argument is that by far the most effective way of reducing drinking problems is by primary prevention at source through a progressive rise in the real price of alcoholic beverages.

The point for present purposes is that the so-called alcoholism epidemic should be seen as a political rather than a medical problem. This has been advocated in an influential article in the *British Medical Journal* by R. E. Kendell, who writes:

The conclusion seems inescapable. Until we stop regarding alcoholism as a disease, and therefore as a problem to be dealt with by the medical profession, and accept it as an essentially political problem, for everyone and for our legislators in particular, we shall never tackle the problem effectively. The medical profession and the caring professions in general are just as incapable of dealing effectively with the harm and suffering caused by alcoholism as the medical services of the Armed Forces are incapable of dealing effectively with the harm and suffering caused by war.[18]

As Kendell also suggests, the disease theory may well be totally at variance with the evidence and a serious obstacle to progress, but its continuing popularity is based on powerful attractions. It is highly convenient for the alcohol-producing industry, when it is criticized for irresponsible advertising or when it is concerned to defend itself against the per capital consumption and restricted-availability kind of argument outlined above. The disease theory allows spokesmen for the alcohol industry to assert that the promotion of its products is directed only at normal drinkers and does not contribute to the number of drinking problems; promotion and advertising cannot affect the few unfortunate individuals who suffer from alcoholism because they were destined to contract the disease anyway; since it is not diseased, drinking by the great majority of people is harmless. So why should this majority be penalized by price increases for the excessive drinking of a few alcoholics? The advantages of this use of disease theory rhetoric became obvious to the American liquor industry very early in the life of the Alcoholism Movement (see p. 46).

The alcohol producers constitute a very powerful political lobby. There is also no doubt that open advocacy of a fiscal preventive policy would be highly unpopular electorally. Perhaps this is why the present government suppressed the Think Tank report (see pp. 307–9) and why its own discussion document, *Drinking sensibly*,[19] comes close to endorsing the same misleading arguments about harmless drinking by the majority advanced by the brewers and distillers. The disease theory is attractive to politicians because it allows them to avoid electorally disadvantageous policy decisions.

Above this, the disease theory has powerful attractions and uses for the whole of society. In an important book entitled *Beyond alcoholism*,[10] Dan Beauchamp has shown how a basic function of the disease theory is to provide an alibi for the drinking of the majority of people and hide the role of alcohol itself in the creation of drinking problems. It does this by 'blaming the victim' and making alcoholism a matter of individual capacities and abilities which the alcoholic lacks. Consider a comparison between the drugs, heroin and alcohol. Heroin is universally regarded as a powerfully addicting substance which anyone could become addicted to if sufficiently exposed; certain individuals may be more likely to try heroin than others but, once addicted, the problem lies not with the addicted individual but with the addictive substance. This was exactly the understanding of alcohol held by nineteenth-century temperance supporters and the majority of those who addressed themselves to 'the drink question'. Modern research has shown that this is indeed the true nature of alcohol and that its only essential difference from a drug like heroin is that much larger quantities must be taken before physical dependence develops.

Yet the disease model, put forward by Alcoholics Anonymous and supported by medical sympathizers, made a fundamental change from this understanding by locating the source of the problem in the vulnerability of the individual alcoholic, and we must ask ourselves why. It did so, says Beauchamp, because, following the Repeal of Prohibition, the demise of temperance ideology and more fundamental changes in social values and aspirations, modern society would not allow alcohol to be blamed for alcohol problems. Hence it was the 'alcoholic' who was blamed. In this way, the rest of society was allowed to proceed with its daily business without being disturbed by any suggestion that the existence of drinking problems was a matter of social responsibility and without being forced to reflect on the implications of its own drinking. The diseased alcoholic was made a scapegoat for society's inability to properly control its favoured psychotropic drug, ethyl alcohol.

Disadvantages for theory and research

This chapter concludes with the most general disadvantages of the disease theory—its drawbacks for our theoretical understanding of drinking problems and the research to which this gives rise. The most obvious disadvantage is simply that the disease theory described in Chapter 3 is massively contradicted by the evidence presented in Chapter 4, and cannot therefore contribute to a valid understanding of what it purports to explain. However, it might be possible to dilute the disease theory sufficiently—perhaps by discarding the assumption of irreversibility, the concept of loss of control, or the postulation of an inherited vulnerability to alcohol—so that it is not so easily refuted. Even if this were done, though, there would remain less specific properties, inherent in calling *anything* a disease, which would continue to retard the progress of theory and research.

In the first place, a disease of alcoholism suggest some kind of unitary phenomenon—a single 'thing' which some people have and some do not. There are several unfortunate aspects of this. The harm caused by excessive drinking is not all-or-none but exists in degrees and it is essential to be able to recognize, describe, and measure these different degrees of harm because they require different kinds of response. The effect of the disease set of assumptions is to distract research attention away from less serious problems or those in the earlier stages of development and to focus concentration almost entirely on the more serious and chronic end of the scale, where it may be too late to do anything effective. By contrast, in drawing attention to *any* harm resulting from excessive alcohol consumption, the learning perspective recognizes the full range of problems which may exist.

A related point is that the assumption of a unitary phenomenon suggests that alcohol-related harm is all roughly of the same kind; it implies that one 'alcoholic' is pretty much like any other. Hence the fruitless search for a single cause of alcoholism in some special biochemical abnormality or unique type of personality. The attraction of this simplistic kind of research endeavour, on which billions of pounds have been spent, is the promise of some magical and dramatic 'cure' for alcoholism, rather like the notorious Keeley's cure at the turn of the century (see p. 40). The truth is that the nature of the drinking problems people experience varies enormously. These different kinds of problem must be studied and understood as phenomena in their own right; their differences cannot be glossed over in an all-embracing disease concept of alcoholism.

The best way of expressing this is to say that a disease concept of alcoholism classifies *people* whereas a more useful perspective invites a classification of *behaviour*. When we think of problem drinking, we should think immediately of the particular behaviour which causes drinking to be recognized as a problem, rather than of a person who is thereby separated from the rest of mankind. At the same time, the dichotomous separation suggested by the disease theory into alcoholics and non-alcoholics is destructive of any attempt to integrate the explanation of abnormal drinking with the wider understanding of normal drinking behaviour, an integration which is essential if genuine theoretical progress is to be made.

This introduces the deepest and most misleading assumption connected with the disease perspective. In suggesting that alcoholism is something a person *has*, rather than something he or she *does*, it insidiously implies that what is wrong is caused by some mysterious process located 'inside' the person. In this way, the occurrence of harmful drinking is explained by the presence of some unknown but confidently asserted physical or psychological malfunction called conveniently 'the disease of alcoholism'. Of course, this 'explanation' merely goes round in circles and, in reality, explains nothing whatever. How do we know when someone is suffering from the disease of alcoholism? When he drinks in a harmful fashion. And what causes this harmful drinking? The disease of alcoholism. This kind of circular thinking is characteristic of pre-scientific modes of reasoning and serves to reassure us that we are making progress in understanding problem drinking when in fact we are not.

The possibility that the disease theory of alcoholism may be representative of a pre-scientific mode of thought brings us to a final point. It should always be recalled that the disease theory only became suspect and an alternative paradigm possible after the introduction of accepted scientific methods to the alcoholism field. It was from the very first employment of such methods—properly designed follow-up studies, controlled laboratory investigations, and painstaking household surveys—that the disease theory began to crumble. It is now clear that the disease perspective was not based on any sound scientific knowledge but on the folk wisdom of alcoholics and their helpers, on hearsay, myth, tradition, rumour, and *ex cathedra* pronouncements of prominent alcoholics and alcohologists.

In the historical context presented in Chapter 2, it was shown how the present-day disease perspective may be traced back to the nineteenth-century moral response to deviant drinking. Thus the disease theory is not only pre-scientific but, indeed, an essentially *moral* conception of problem drinking. This nineteenth-century disease model may well have been essential in the 1930s to engender compassion and help for self-confessed alcoholics. Since then, however, behavioural psychology has made great strides and has succeeded in transforming itself from a largely mechanistic and individualistic model of behaviour to a fully cognitive and social account of human life. Thus, an alternative explanation of problem drinking, an alternative ground for compassion and care, is now available.

The disease theory has outlined its usefulness. It is time to move on to a genuinely scientific, empirically based paradigm of problem drinking.

Part 2

Problem drinking:
a social psychological paradigm

6 Setting the scene

Our task in the second part of this book is to present a credible alternative to the disease perspective through which to understand and respond to problem drinking. This chapter sets the scene for the three remaining in Part 2. This is achieved by an attempt to define the nature of problem drinking and by an outline of the approach known as *social-learning theory* which will serve as the basis for the alternative paradigm of problem drinking to be described later. We begin, however, with two thoroughly typical case histories.

Two case histories

Case one

The patient, Colin M., appears at the clinic in a nervous, depressed, and dishevelled state. He has been abstinent for three days but reports having constantly to resist craving and has almost relented on two occasions. On one of these he actually got to the door of his local, changing his mind only at the last minute. He is extremely guilty about his past behaviour and very frightened that his wife will leave him. His self-esteem is low and he feels hopeless about overcoming his problem, since he has tried to stop or cut down at least a dozen times before.

Colin's job is under threat and he is on a last warning for bad timekeeping, inefficiency, and absenteeism. He is in considerable financial difficulties and has large debts. Physically, his health is poor and he is currently being investigated for a suspected duodenal ulcer. His wife has left him previously on three occasions because of verbal, and occasional physical, abuse and he attends this clinic under threat of her leaving him for good. The two children, Carol aged 3 and John aged 7, are somewhat alienated from their father and John is attending the paediatric clinic because of persistent nocturnal bed-wetting.

On interview, Colin's wife Sheila appears stressed and is tearful at times. She complains about the persistent arguments at home and is extremely worried about the possibility of Colin losing his job. They find it difficult to talk calmly together without it escalating into a major row and their relationship, sexual as well as emotional, is at an all-time low. Sheila reports that Colin had managed to cut down after they married seven years ago and she thought she had changed him for the better. Over the last two years, however, Colin has tended to disappear more and more frequently and the tension at home has steadily increased.

Colin confirms this account and adds that over this period he has gradually become more isolated from his friends and has dropped his old pastimes and sports, such as badminton, which he used to play every week. He reports losing interest in most other things including, to his guilt and shame, his family. He knows that he has changed and feels like a Dr Jekyll and Mr Hyde character, with Mr Hyde coming to dominate more and more as the problem progresses. What bothers him most is that he does not even enjoy it any more; he feels driven to carry on even though he feels guilty and depressed while doing so. He wakes up in the morning feeling a sense of dread at the prospect of facing up to another day with a mind so preoccupied. Sometimes he even finds himself trembling and drenched in sweat on wakening and his sleep is very disturbed. Recently things got so bad he thought about killing himself. At present he is quite desperate and does not know what to do.

Case two

Mary D. has more than quadrupled her consumption over the last year, and now finds herself bingeing even in the middle of the night and at work where she keeps supplies in her office desk. She presents as an extremely depressed, 33-year-old woman whose marriage has just broken up. The separation was, in her opinion, attributable in part to her excessive consumption over the last year. Mary complains primarily of a loss of control over intake which results in frequent binges. The associated guilt and depression leads to her spending whole days in bed when she should be at work.

Mary experiences particularly intense craving in the evening when she is alone, but the desire is now coming on her more often at other times. She finds herself planning ahead to ensure an adequate supply and, if she doubts whether enough will be available on a given occasion, she takes her own additional supply in her handbag. If she takes just one, she feels an irresistible urge to continue, and this 'avalanche' can lead to a long binge, with the inevitable after-effects of nausea, vomiting, depression, guilt, and self-disgust. The violent arguments with her husband which surrounded these binges exacerbated her mental distress and made subsequent binges more likely.

The steady increase in consumption over the last year has been punctuated by periods of restraint, usually lasting two weeks at most. Each relapse seems worse than the last and Mary appears to have lost all confidence in her ability to control her intake. Her husband accuses her of being weak-willed and a bad wife, and has not offered much support. She feels as if she is the victim of some overwhelming compulsion over which she has little or no control and is quite desperate for help.

Colin and Mary do not drink. Colin is a compulsive gambler and Mary a compulsive overeater. Yet their stories are similar to those of many problem drinkers. The reason for describing a gambler and an overeater in a book about problem drinking is this very fact, that the parallels between the problems are so great. Perhaps by looking at disorders which are relatively free from issues of drug and illness, we can see more clearly some of the central features of problem drinking. Alcohol is, of course, a drug, but are its pharmacological properties the major determinants of its misuse? While they are important, relatively more important is the psycho-social context in which these pharmacological effects are experienced by the individual. In other words, these effects are, to a great extent, *learned*. Furthermore, in so far as this is true, they should, in theory at least, be capable of being *unlearned*.

Too much of a good thing

The last few years have seen a proliferation of track-suited runners, showing greater or lesser signs of cardiac distress, pounding the streets of our towns and cities. While there appears always to be a medical expert ready to make a professional reputation by immortalizing a piece of faddish human behaviour in the jargon of a syndrome, reports of the existence of 'jogging addicts' may be of some relevance to the discussion in this chapter.

These 'addicts' are reported to have suffered broken marriages because of an obsession with running, to persist in their behaviour in the face of physical damage, such as broken bones and strained ligaments, and to show an increased tolerance for exercise in the sense that, whereas three miles an evening satisfied them before, they now need a six-mile run to feel properly relaxed. A certain restlessness and moodiness is also apparent if they cannot, for whatever reason, complete their daily run.

It is almost a defining characteristic of the human species that it can demonstrate apparently self-damaging behaviour. This usually occurs when the particular behaviour offers pleasurable immediate consequences, yet at the same time incurs damaging long-term consequences for the individual. To name only some of these two-edged swords: smoking; various forms of sexual behaviour; gambling; many types of drug intake; dangerous sports (motor racing, for example); voyeurism; exhibitionism; some forms of criminal behaviour; overeating.

What grounds have we for suggesting that problem drinking should be classed as a close relative of these phenomena? To justify this assertion, we will examine the defining features of the 'alcohol-dependence syndrome'. As noted in Chapter 3 (pp. 84–7), this has become established as the term to replace 'alcoholism' in the International Classification of Diseases and has been of great importance in the last few years in shaping thinking and policy about problem drinking.

In 1977, Griffith Edwards and his colleagues[1] described the following seven elements of the alcohol-dependence syndrome:

(1) narrowing of the drinking repertoire;

(2) salience of drink-seeking behaviour;

(3) increased tolerance to alcohol;

(4) repeated withdrawal symptoms;

(5) subjective awareness of a compulsion to drink;

(6) relief or avoidance of withdrawal symptoms by further drinking;

(7) reinstatement of dependence after abstinence.

We shall examine each element in turn and consider the extent to which they apply to other forms of excessive behaviour.

Narrowing of the drinking repertoire. By this, Edwards and his colleagues refer to the way the problem drinker's drinking becomes increasingly stereotyped. Instead of moderating consumption according to mood, circumstance, and responsibilities, problem drinkers drift towards a state where alcohol is like a vital fuel, without which they fear the engine of their lives will grind to a halt. They tend to drink as much on Monday night as on Saturday night or to take in as much before their children's school concert as before the staff dance.

Such a narrowing of behaviour is not confined, however, to the drinker who is becoming dependent on alcohol. The gambler whose predicament was described at the beginning of this chapter also found that his gambling behaviour was becoming increasingly 'stereotyped'. Rather than enjoy a flutter on a Saturday morning and forget it on Tuesday night when money runs short, he begins to borrow, beg, or steal money so that he can maintain his habit throughout the week. Instead of gambling fitting into his life, Colin begins to arrange his life around the habit, at the cost of his marriage, job, health, and finances. Priorities shift and Colin's gambling becomes increasingly predictable and narrow in its manifestation.

The overeater Mary also shows a similar pattern: chocolates are nibbled at the office desk; furtive midnight forays to the refrigerator extend the eating habit far into the night; the amount eaten on any one occasion becomes increasingly inflexible; and thoughts of food spread more and more into consciousness. Similar encroachments of their habits into the lives of smokers, heroin addicts, and even urban joggers may be observed, and the processes by which this narrowing of behaviour occurs will be discussed in the following two chapters.

Salience of drink-seeking behaviour. Closely related to narrowing is 'salience'. Simply stated, this means that the person devotes a lot of time, effort, and thought to ensuring an adequate supply of the substance or activity. The gambler is preoccupied with loans, deals, permutations, tips, and all the other accoutrements of the habit; long hours are spent studying form or standing in casinos and working fruit machines. The exhibitionist may spend considerable time pacing the streets looking for a suitable opportunity to 'flash'. The smoker will scour the sleeping city for a cigarette machine if the supply of cigarettes seems theatened. The compulsive shoplifter may wander round big stores for many hours preparing for the theft, while the youth compelled to steal cars for joyrides spends his evening sorting and labelling hundreds of car keys. The old coffee-hand will break the most demanding routine to secure his 'fix' of caffeine. Whenever someone wants to do something very badly—however ambivalent he is about doing it—preparation for that activity will be salient in thought and/or action.

Increased tolerance to alcohol. Tolerance is usually applied to drug effects and refers to the phenomenon whereby increasing doses of a drug are required to produce the same effect. Almost everyone who drinks alcohol will have experienced this, in so far as the first couple of drinks ever taken had a stronger intoxicating effect than the same drinks taken a few years later. Pain-killers taken regularly usually lose some of their potency unless the dose is increased, and the insidious increase in quantities required by the heroin addict is a familiar story.

Is tolerance exclusively in the province of pharmacology? In the next chapter we will describe how tolerance to powerful drugs can, in fact, be learned and unlearned and is thus, at least partially, a psychological phenomenon. It is not surprising therefore to observe tolerance in other spheres of life than drug-taking. Most compulsive gamblers will report how the one-pound bet gradually lost its power to excite and had to be progressively increased; some sexual fetishists are impelled towards increasingly bizarre activities to sustain the erotic 'kicks'; and the panting jogger, notching up his daily route, mile by painful mile, is no longer satisfied with the modest two miles achieved in the first week. It would be stretching the argument to propose that the development of tolerance to these varied activities is exactly the same sort of process as the development of tolerance to alcohol and other drugs, but the similarities are there and cannot be ignored.

Repeated withdrawal symptoms. Psychologists Ian Wray and Mark Dickerson,[2] formerly of the Royal Edinburgh Hospital, reported the following symptoms among a group of patients who had abruptly abstained; 'I felt irritable'; 'I felt restless'; 'I felt fed-up (depressed)'; 'I could not concentrate'; 'I kept thinking the same thoughts over and over again'; 'I felt anxious'; 'I felt jumpy'; 'I felt angry'; 'I felt bored'.

These symptoms make up a kind of withdrawal syndrome. In this case, however, the withdrawal was not from alcohol but gambling. The respondents were all compulsive gamblers but the symptoms they experienced are all commonly reported by problem drinkers who stop drinking suddenly. While it is obvious that there are many largely pharmacological aspects to alcohol-withdrawal symptoms, such as whole-body tremor, at least some of these symptoms are, to a significant extent, psychological phenomena which transcend particular drug effects and are common to a number of non-drug withdrawal states.

Many of these withdrawal symptoms are also symptoms of general distress. Sweating, restlessness, fear, and despair on awakening, tremor, and nausea are all associated with anxiety and are common reactions to the loss of valued objects or people. That someone who has been immersed in a particular drug or activity for many years should experience such symptoms of distress when they stop the habit is hardly surprising.

The extent to which these symptoms are attributable to pharmacological factors or to psychological processes will depend on the nature of the addictive behaviour. However, a strong, common, psychological thread runs through the withdrawal states produced by the cessation of all addictive patterns. Heroin addicts experience much greater physiological changes during withdrawal than compulsive gamblers, but the type of effects on a person's psychological state from the loss of an important activity may be similar in both cases. In the next two chapters we shall outline just how closely the physiological and psychological systems involved in the addictive behaviours are interlinked.

Subjective awareness of a compulsion to drink. Central to the puzzle of addiction is the fact that addicts often do things which, apparently, they neither intend nor want to do; in other words, their control over drinking is, in some sense of the word, impaired. However, disease models of problem drinking have proved wanting in their physiological explanations of impaired control. As we have seen, a fundamental problem with such explanations is their inherent circularity: heavy drinking is explained in terms of impaired control, yet the existence of impaired control is inferred from the presence of heavy drinking. In other words, the concept of impaired control is tautologous and conceptually indistinguishable from what it purports to explain. As a descriptive device, however, impaired control may be of some value as shorthand for a number of phenomenological and behavioural manifestations of the addictive cycle.

The compulsive eater who crams chocolate after chocolate into his mouth until the supply runs out or nausea overcomes him is demonstrating impaired control over his behaviour. The 'workaholic' who works into the night, missing her dinner engagement, doing that 'just one more' piece of work, is also showing some kind of impaired control, as is the runner who drives himself to exhaustion and injury by pushing on beyond his physical limitations. It is the paradox of such phenomena that the person's behaviour is at odds with his or her enjoyment of the activity, even though there are no obvious external spurs to such frenetic endeavours. This is the compulsive element to the addictive cycle and is evident in the gambler who feels driven and miserable as she places a bet she cannot afford, and in the drinker who gets slowly more depressed as the whisky bottle empties.

It is this divergence of experience and behaviour which drove theorists to insist that something physical must account for the unruly autonomy of addictive behaviour. We shall show in the next two chapters how recourse to such explanations is not necessary and how the germs of an explanation of 'impaired control' exist in well-established psychological theories.

Relief or avoidance of withdrawal symptoms by further drinking. It is perhaps getting nearer to the essence of addictive behaviour to talk of doing something to avoid feeling worse, as opposed to doing it to feel better. Seen in these terms, the disparity between behaviour and expressed wishes becomes less puzzling, because it is largely the result of being made 'an offer you can't refuse'. This leads to the kind of unease and restlessness felt by recipients of similar offers in gangster films, where accepting the offer means experiencing consequences only marginally less punishing than those which would follow refusal. In this situation, ambivalence, self-castigation, anxiety and low mood can be expected, and this is exactly what happens to addicts who do what they do because of what they fear will happen if they do not.

This phenomenon of 'negative motivation' is not the exclusive domain of the pharmacological addictions. The distress which follows the loss of a familiar activity is a common spur to the resumption of that activity. For example, the gambler places bets as a means of reducing restlessness and tension; before they steal, some compulsive thieves report a build-up of unpleasant tension, which the theft seems to dissipate; many compulsive eaters report that their eating at times constitutes an attempt to dispel bad moods and tension.

People tend to become addicted to activities or substances which produce, in the beginning at least, pleasant and fairly immediate changes in their psycho-physiological state. Most people feel somewhat 'flat' after an intensely enjoyable experience, especially if they do not have some satisfying alternative activity in which to immerse themselves. If the activity or drug is taken many times in close succession, feeling flat can come to be synonymous with the absence of the drug or activity. Conversely, feeling 'high' or good becomes linked with the addictive behaviour. Unfortunately, as mentioned in the discussion of tolerance, most activities or drugs which produce pleasant effects gradually lose their power to cause these changes. The human brain is programmed to react to unfamiliar events and, in order to do so, 'damps down' its reaction to events which are repeated again and again.

We are left with a gradually waning effect of the addictive behaviour, in the context of a continuing expectation that the effect will persist. The failure of this expectation to be met may lead to a perpetuation of the flatness in the absence of the drug/ activity. And in the absence of strong pleasant effects, motivation for taking the drug or engaging in the activity may change imperceptibly to one of trying to avoid the flatness, which has come to be associated with not having the drug/activity.

We have anticipated the material to be discussed in the next two chapters, where much of the argument will become clearer. The purpose of introducing this explanation here is to emphasize the psychological components of addiction and to show that relief consumption is not an exclusive property of ethyl alcohol. On the contrary, doing things to avoid feeling bad is a common human experience which can lead to many problems— in relationships, sports, crime, sex, and also drug-taking.

Reinstatement of dependence after abstinence. This aspect of dependence is perhaps the most difficult from which to extract psychological commonalities. It refers to the phenomenon whereby symptoms of dependence on alcohol, which have developed over many years of heavy drinking, re-emerge quickly once a person who has abstained for a time starts to drink again. It appears as if the brain is 'sensitized' to the drug and that the person quickly returns to square one, in spite of having achieved a spell of abstinence.

It is outside the scope of this chapter to examine the evidence and attempted explanations for this phenomenon, but it is possible to make some relevant observations about drug-free addictive behaviours.

Consider an overeater who has gradually adopted bad eating habits over a number of years but who manages to go on a strict diet for several months. What is the likely reaction of this person to eating some rich chocolate cake? Some dieters might use this as a spur to embark on an even more Spartan regime (and in some serious eating problems, such as anorexia nervosa and bulimia, this is especially true), but many would experience guilt, depression, hopelessness, anger, resignation, anxiety, and despair. The chances of the poor ex-dieter shrugging her shoulders and saying, 'I might as well be hanged for a sheep as for a lamb' are high, and this thought might well trigger off an escalation into binge-type eating. The resulting orgy would be more likely to resemble the most recent style of overeating than that which preceded the obesity. In other words, we are observing a rapid reinstatement of a behaviour pattern which took years to build up. This may be because thoughts and emotions evoked by the 'slip' are more similar to those associated with the most recent binge than to earlier, more accept-abled eating patterns. Similarly, when the problem drinker slips, he experiences a flood of memories, thoughts, and emotions unleashed by association with the act and context of drinking. Such a barrage of stimuli is more likely to set off the behaviour with which the stimuli are most strongly associated, that is, heavy drinking, than other behaviours, such as abstaining or drinking moderately. Included in this reaction are the emotions of anxiety and depression which are part of the withdrawal syndrome.

It would not be difficult to give an example of reinstatement from other addictive behaviours like gambling. There is no doubt, however, that there is a strong drug effect which complicates the picture in drug-related addictions. Anyone seeing a problem drinker start to drink heavily again following a period of abstinence would scarcely claim that the resulting physical debility could be attributed to purely psychological processes. What could be claimed is that at least some of these physical reactions are conditioned—that is, produced by association with certain events such as the presence of alcohol in the blood. This is not to suggest that these physical reactions are in some way imagined or 'put on'. The proposal is that very real physiological responses are evoked by internal or external stimuli and, furthermore, that this learned connection might be amenable to manipulation by psychological means. We shall go into these issues more fully in the next chapter.

So far in this chapter, we have tried to 'take the drug out of addiction' and to emphasize the common psychological processes apparent in a wide range of addictive behaviours, including those labelled as harmful, abnormal, or deviant and those regarded as socially acceptable. Placing addiction squarely within the sphere of normal human behaviour has a number of advantages, not least of which is that 'normal' people are thereby discouraged from keeping 'alcoholics' or 'addicts' at arm's length by classifying them as oddities whose behaviour is unintelligible. Another advantage is that we can now bring to bear the large body of knowledge about normal human behaviour to assist an understanding of the addictions; later in this chapter we shall outline the origins of an approach to human behaviour which may serve this purpose.

Before moving on, we should make it clear that the argument just advanced for the existence of commonalities in the 'addictive behaviours' is quite separate from the craze in the United States for labelling all kinds of behaviours as 'addictions', which was mentioned in Chapter 2 (p. 60). In the latter case, the attempt is made to extend the concept of alcohol addiction as a disease to a whole range of diverse forms of behaviour and to argue that these can only be 'cured' by quasi-religious methods similar to those of Alcoholics Anonymous. Ironically, modern psychological theory supports the idea that dependence on drugs and dependence on non-drug forms of compulsive behaviour have much in common. However, this is used to argue that alcoholism and other forms of drug dependence are *not* usefully seen as diseases. Thus the implications are exactly the opposite from those used to buttress the profit-making treatment ventures in the United States.

What is problem drinking?

In an eagerness to dissolve the boundaries between normal and abnormal drinking and get away from a disease-laden 'them and us' conception of problem drinking, we may have appeared to minimize the misery and suffering of problem drinkers and their families. Our intention could not be further from this; to deny that problem drinking is a disease in no way diminishes a plea for compassion, understanding, support, and help for those affected. Many problem drinkers show some or all of the following problems: they develop serious physical illnesses; they commit suicide; they experience severe depression and anxiety; they suffer prosecutions, evictions, and multiple legal suits; their marriages split up; they become isolated and friendless; they develop low self-esteem; and they die younger than they should.

But these palpable facts do not help to define problem

drinking. One possibility is to borrow Jellinek's wide-ranging definition of alcoholism (see p. 7): 'any use of alcoholic beverages that causes any damage to the individual or to society or both'. This kind of all-embracing definition has advantages and disadvantages. Among the advantages is that it encourages us to pay more attention to the less severely disabled problem drinker, whose prognosis is so much more favourable than that of the classic 'alcoholic'. The main disadvantage is that almost every drinker could fall into this new definitional net! Few drinkers have *never* had a hangover, *never* said the wrong thing after a few drinks, or *never* felt a little sluggish at work owing to the effects of alcohol.

Clearly, the definition of what constitutes a drinking problem is intimately bound up with who is making the definition, as well as with the nature of the social group among whom problems are being defined. What may be a problem to a white-collar worker's family may be seen as relatively normal behaviour among a manual worker's; drunkenness in a woman may be regarded as a serious problem, yet similar behaviour in a man may be seen as a display of healthy masculinity. Official concern with alcohol misuse grew with industrialization. When ordinary people worked in the fields and lived the life of subsistence farmers, drunkenness did not seriously interfere with their way of life. Once factories sprang up, the havoc wrought by alcohol on the disciplines of work and time-keeping brought drink to the top of the agenda for factory owners and for governments.

Nevertheless, some consensus about the nature of problem drinking in our society is possible. The first point is that any useful definition must include the notion of *repeated use* leading to some kind of harm. The types of harm possible have been neatly categorized by American sociologist Ron Roizen[3] into the '4 Ls'—liver, lover, livelihood, law. 'Liver' is shorthand for physical problems of all kinds, and also the psychological distress, such as depression and anxiety, which can arise from heavy drinking. 'Lover' stands for problems of interpersonal relationships, whether these involve family or friends. 'Livelihood' refers to problems of employment, from being demoted to being sacked, and encompasses educational failures and difficulties, financial problems, and other consequences. Finally, 'law' represents the various civil and criminal proceedings which may arise from alcohol misuse.

Different aspects of drinking cause varying types of problem. For instance, the family of a Hampshire squire may see his health deteriorate owing to regular daily libations of gin, claret, and port, yet there may be no family, legal, or financial problems. In this case, it is the first 'L'—the liver—which suffers, and the other three are unaffected. A Tyneside shipyard worker, on the other hand, may run into severe marital problems because of the sixteen pints of ale he drinks on a Friday and Saturday night. He may or may not have problems in other areas, but it is the second 'L'—lover—which is most obviously affected. The Glasgow factory worker whose husband is unemployed, and who is supporting the family on her meagre wage, may risk losing her job because of the lunchtime vodkas she drinks; her efficiency at a job which demands precision and manual dexterity may be impaired. In her case it is livelihood, the third 'L', which is principally at risk. The travelling salesman who treats potential customers to frequent 'liquid lunches' and who often drives with more than the legal limit in his blood may lose his driving licence. Thus legal problems, the fourth 'L', may adversely affect his career. In many cases, all four 'Ls' will suffer, but one must avoid the danger of creating stereotypes, and individual differences in the types of problem which can arise must always be emphasized.

An alternative way of classifying alcohol problems has been devised by psychiatrist Anthony Thorley.[4] This classification is based, not so much on the aspect of a problem drinker's life which may be harmed, as in Roizen's '4 Ls' classification, but on the aspect of drinking which causes the problem in the first place. Thus, Thorley identifies three types of drinking problem: those arising (1) from regular excessive consumption; (2) from intoxication; and (3) from alcoholic dependence (see Fig. 6.1 p. 193).

The first of these relates to chronic intake of alcohol, that is, how much is drunk over weeks, months, and years. Typically, this type of problem involves health (for instance, liver cirrhosis or pancreatitis) and/or financial difficulties, though legal and livelihood problems may also emerge. While there are wide individual variations in tolerance to alcohol and while some people more easily suffer ill-effects at a given level, a weekly average consumption of about fifty units of alcohol per week for men (one unit equals one half-pint of beer, one measure of spirits, or one glass of wine) seems to represent a turning-point which, if exceeded, results in a disproportionate increase in certain drinking symptoms, such as alcohol-related tremor, as well as certain physical ailments.[5] For women, the evidence about risk levels is scanty, and a given amount can lead to different concentrations of alcohol in the blood at different times in the menstrual cycle. While drawing arbitrary lines is always a tendentious business, a consensus is beginning to emerge that the *upper* weekly limits should be thirty-five units for men, and around twenty for women.

It is possible, though, to drink less than these weekly totals and still be drinking too much. The most obvious example is where large quantities are consumed in short periods of time, which brings us to the second dimension of problem drinking— problems due to intoxication. Apart from the health, social, legal, and employment implications of such drinking patterns, rapid drinking more readily leads to blackouts (amnesias for what has happened while drunk), as well as to passing out (losing consciousness). Obviously, accidents occur as a result of intoxication and, in Scotland in 1978 for instance, 49 per cent of car drivers and 67 per cent of pedestrians killed on the roads had blood alcohol levels in excess of 80 mg per 100 ml.[6]

Family violence, trouble with the police, assaults, and many other consequences also arise from intoxication and it is this type of drinking problem which is common in northern European countries, such as Scotland, Sweden, Finland, and the Soviet Union. In the southern European countries—the so-called 'wine cultures'—much more is drunk but it is spread out over longer periods, so that physical illness, and road and industrial accidents emerge as the major kinds of problem in countries such as France and Spain.

The third dimension of drinking problems, those arising from alcohol dependence, should not be confused with a disease model of alcoholism. Dependence is an essentially psychological concept (see pp. 178–88), which nevertheless encompasses a number of physiological events, such as tremor, sweating, and nausea, that are incorporated into the learning processes leading to dependence. We shall not try to define dependence here, since the next two chapters can be seen, in one sense, as an attempt to do so.

The relationships between the three aspects of a definition of problem drinking are presented schematically in Fig. 6.1. A problem drinker is thus defined as anyone falling within the perimeter of the three circles; a person can be designated a problem drinker by showing one, two, or all three elements of the scheme. Clearly, a drinker located at the intersection of the three circles has a severe problem, and it is within this group that the stereotypical 'alcoholic' would be located.

An alternative set of assumptions

Having attempted to define problem drinking, let us now list the core assumptions of the paradigm on which the new understanding is to be based. At the end of Chapter 3, four key assumptions of the disease theory were isolated (pp. 88–9): namely, that alcoholism is a discrete entity, that alcoholic drinking arises because of craving and loss of control, that the

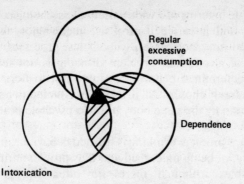

Fig. 6.1. Three aspects of drinking which can cause problems [adapted from Thorley, A. (1985). The limitations of the alcohol-dependence syndrome in multi-disciplinary service development. In *The misuse of alcohol: crucial issues in dependence, treatment and prevention* (ed. N. Heather, I. Robertson, and P. Davies). Croom Helm, London].

latter are irreversible, and that continued drinking by an alcoholic invariably leads to a progressive deterioration. These assumptions were reviewed in the light of the evidence presented in Chapter 4 (pp. 140–1). Following this review and the further discussion in Chapters 5 and 6, it is already possible to identify an alternative set of assumptions applying to the more detailed paradigm of the next three chapters:

1. Problem drinking is not a discrete entity; there is no hard and fast line separating problem drinkers from non-problem drinkers. At the same time, problem drinking is not a unified phenomenon. Rather, it is learned behaviour which is nested in complex ways within the socio-psychological world of each individual.

2. Problem drinking cannot be explained away by tautologous concepts like 'loss of control' and 'craving'. Rather, it is

intelligible in terms of a wide variety of psychological processes. Craving and impaired control are meaningless terms unless translated into specifiable psychological processes. Unlike the disease of alcoholism, problem drinking is not seen as completely determined behaviour, in contrast to normal drinking which is freely chosen. All drinking is assumed to be capable of explanation by the same body of socio-psychological principles.

3. Since problem drinking is learned behaviour, it is capable, in theory, of being unlearned and thus problem drinking is not irreversible. Although there are often strong, pragmatic grounds for total abstinence as a solution to a serious drinking problem, it is possible for many other problem drinkers to return to a controlled pattern of use.

4. There is no inexorable sequence of events in problem drinking, and continued drinking by a problem drinker does not necessarily result in a progressive deterioration. Treatment is not always essential as a solution to a drinking problem, although some problem drinkers do get worse over time and do require specialized help.

Social-learning theory

Man is not a completely rational creature, guided purely by cool reflection and enlightened contemplation of the world. *Homo sapiens* is a motley assembly of primitive responses and superbly sophisticated intelligence. As a species, we can build machines which fly to the moon, yet be paralysed with fear at the sight of a small brown rat. We can talk rationally about the possible end of life on earth, yet behave in such a way as to make it more likely. There are still chest surgeons who smoke and priests who go to war.

People behave the way they do largely because they *learn* to behave in that way. Of course, genes also have an effect, but genetic tendencies are to a great extent *plastic*; that is, the way they are expressed is shaped by experience and learning. For instance, man has the genetic capacity to be aggressive, yet some cultures foster the expression of aggression more than others.

Learning can take place on a number of levels, and social-learning theory is the systematic study of the way humans learn to behave, think, and feel. It differs from attempts to explain human behaviour like psychoanalysis because it uses the scientific method, with its emphasis on repeatable measurement of the processes studied. It also differs from psychoanalysis because it recognizes a number of different levels of learning, with conditioning at one extreme and abstract thought and reflection at the other. Each of these levels has been experimentally studied so that there now exists a large body of information which allows us to make predictions about how people behave under certain conditions.

More primitive processes, such as conditioning, are not incompatible with high-level thought but their relationship can be, at times, an uneasy one. A spider phobic may agree that there is no rational basis for his fear, yet quake at the sight of the spider. Thus, while man can be rational and reflective in his actions, the fact remains that he is also sometimes at the mercy of lower-order mechanisms of learning. The relationship between these various levels is as yet far from clear and social-learning theory is still patchy. Yet there is much of value in helping us to understand some aspects of human behaviour and experience. A somewhat arbitrary division of these levels is: classical conditioning; instrumental learning; modelling; self-management/self-regulation; higher cognitive processes.

The order in which these levels are presented corresponds roughly to the order of their historical appearance in this century. We will outline briefly the research findings on each and their development in the historical context.

Classical conditioning

Ivan Pavlov, director of the physiological laboratory at the Institute of Experimental Medicine in Moscow, was awarded the Nobel Prize in 1904 for his work on glandular and neural factors in digestion. Ironically, he was not destined to become famous for this, but for the accidental discovery of a fundamental psychological process known as the conditioned reflex. Doubly ironic is the fact that Pavlov was a reluctant psychologist and was not enthusiastic about the emerging discipline.[7] While he is certainly not the only researcher whose work demonstrated the existence of conditioning, we shall use Pavlov's famous studies as an example.

The accidental discovery was made while Pavlov was measuring salivation by dogs receiving food. He noticed that the dogs would salivate in anticipation of food—for example, when they caught sight of the attendant approaching or saw the food dish. This led to an extensive research programme which systematically examined the conditions under which connections between certain stimuli (the sight of the food dish) and certain responses (salivation) were made. Pavlov demonstrated, for instance, that if a bell were rung whenever meat powder was given to a dog, the bell alone could eventually make it salivate when previously only food could have had that effect.

This discovery was one of the most important in psychological science because it allowed researchers to quantify the processes observed and, hence, systematically investigate the laws governing some vital psychological functions. A number of conclusions about conditioning emerged from Pavlov's work and that of subsequent researchers, and we shall outline the most important. Before this, however, let us illustrate how the basic conditioned response is acquired.

First, an unconditioned stimulus exists (a loud noise, say) which produces an unconditional response (a start or fright):

A.   Unconditioned stimulus   →   Unconditioned response
                (noise)   (start)

Then a neutral stimulus (for example, a red light flashing) is
presented shortly before each unconditioned stimulus:

B.   Neutral stimulus   paired with   Unconditioned stimulus
            (red light)   (loud noise)

After this has happened several times, the neutral stimulus
comes to have an effect similar to the unconditioned stimulus;
it produces a similar response. When this occurs, the neutral
stimulus has become a conditioned stimulus and then evokes a
conditioned response very like the unconditioned response.

C.   Conditioned stimulus   →   Conditioned response
                (red light)   (start)

We now have the situation where a person gives a start
whenever a red light flashes, even though no loud noise
follows. It should now be clear that conditioning plays an
important part in human behaviour. For instance, sexual
arousal is conditioned in many people to a variety of stimuli
other than those directly involved in sex. Certain types of
night attire are marketed on the assumption that they arouse
the consumer and her companion; some people are 'turned
on' by bizarre objects which have no obvious connection with
sex—rubber clothes, whips, leather, and so on. In most cases,
these have developed because the object of the fetish has been
associated with sexual arousal and, hence, has become conditioned to it. To turn to a more mundane example, the smell
of newly baked bread is likely to cause our mouths to water
and make us feel hungry. Similarly, simply going into a
kitchen can produce hunger, as can the sight of clock hands
approaching lunch hour. Even the meekest dog can terrify the

child who has been bitten by another dog just once, and the man who has been stuck in a lift on a single occasion may panic whenever he goes into one.

These are all examples of conditioned responses which play an important part in our daily life. In evolutionary terms, they have been vital for survival because they have enabled us to react quickly to threatening situations and alert us to the presence of such necessities as food and sexual partners. In psychological terms, they save us from information overload by delegating to automatic systems a set of responses which, if each had to be scrutinized and rationally chosen, would overwhelm the limited capacities of the brain. Of course, these automatic systems can run awry and the development of sexual fetishes, phobias, compulsive eating, and many other problems bears testimony to this. But this does not detract from the fact that they are most important for our survival.

As will be obvious, human behaviour cannot be explained solely in terms of conditioned responses; humans think, plan, believe, expect, and anticipate; they influence the environment as much as it influences them; they are certainly not passive responders to conditioned stimuli. Modern social-learning theory attempts to elucidate the principles governing these processes, just as it attempts to explain the way in which conditioned responses develop. Conditioning and thought are not incompatible; they represent different levels of learning and reciprocally influence each other. Conditioned responses can be made and unmade by thought, while thought can be influenced by conditioning.

This brief review cannot do justice to a number of other important principles of conditioning, such as generalization, extinction, incubation, and spontaneous recovery. In the next chapter we shall outline the relevance of some of these processes to problem drinking.

Instrumental learning

It is a consequence of the existence of residual, primitive circuitry in the biological make-up of human beings that they tend repeatedly to do things which give them immediate pleasure. Conversely, they tend not to do things which deprive them of satisfaction or inflict punishment. In evolutionary terms, this makes admirable sense; sex is pleasurable because it drives us to reproduce, and eating because it stops us from starving to death.

When the human race has developed the intellectual capacity to destroy the entire planet, however, one can see limits to the survival value of this primitive mechanism. Unfortunately, it is the strength of the mechanism which is one factor in keeping us polluting the atmosphere, eating harmful foods and, generally, engaging in activites which give, in the short term, individual reward but in the long term hurt both individual and society. Thus, co-existing in human beings is a conundrum of primitive impulse and elevated reflection. Instrumental learning is the study of an important but primitive psychological mechanism which operates with varying degrees of independence from higher-level psychological processes.

The origins of the instrument-learning approach lie in the work of E. L. Thorndike, a psychologist at Columbia University, New York, who studied how animals learn to find their way through mazes, escape from 'puzzle boxes', and solve a number of related problems. What struck Thorndike forcefully during these experiments was the trial-and-error quality of learning, and the way accidental successes led to the strengthening of responses which preceded each success. Thus, a cat would paw and nuzzle its way around a cage in random fashion until one particular action had some effect, such as causing one part of a door-catch to loosen. The cat would repeat this action until something else happened, and so on. In 1905, Thorndike formulated the simple but crucially important *Law of Effect*:

'Any act which in a given situation produces satisfaction becomes associated with that situation, so that when the situation recurs the act is more likely than before to recur also. Conversely, any act which in a given situation produces discomfort becomes disassociated from the situation, so that when the situation recurs the act is less likely than before to recur'.[8]

The study of instrumental learning burgeoned during this century in the USA. The most famous exponent of operant conditioning, B. F. Skinner, asserted that psychology was at too early a stage in its development as a science for theorizing to be anything more than a distraction and a handicap to clear observation and experiment. He adopted a purely empirical, atheoretical approach to the study of how reinforcement—his term for Thorndike's 'effect'—influenced behaviour, using the so-called 'Skinner boxes', in which pigeons and other animals could receive automatically dispensed food for performing predetermined sequences of actions.

Most operant-conditioning research has been carried out with animals, though in recent times some of the findings from animal research have been replicated with human beings. In one study,[9] subjects were wired up to a machine which could detect tiny, imperceptible muscle contractions beneath the skin, not noticeable to anyone including the subjects themselves. Only the researcher could see whether the muscles were contracting by means of amplification on a screen. Whenever they did so, the subject was given money as a reward and, as a result, contractions increased considerably even though the subjects did not know what they were being rewarded for. This is clear evidence that instrumental learning can take place, under some conditions, without awareness.

Modelling

In a letter to his son in 1750, Lord Chesterfield wrote: 'We are, in truth, more than half what we are by imitation.' A newspaper strike in a mid-western city of the USA in the late 1960s was followed by a slump in the suicide rate of that city; athletes run faster when running against others than against the clock; studies of prejudice in the United States have shown that it is common for whites who have never met one to show extreme prejudice towards blacks.

What is the common feature of these three very different examples? The answer is *modelling*; in each case, learning has taken place through observing the behaviour of others. It is a well-documented fact throughout the world that a dramatic and well-publicized suicide tends to be followed by a spate of similar suicides; hence the care usually taken by the media not to give wide coverage to such events. Athletes are spurred on by the presence of high-performing adversaries, and racialist and other prejudices are often acquired by observing the prejudices of others, usually parents.

Why do we imitate others so often? The answer may be that, in evolutionary terms, doing as others do has been the best way of surviving. As a young member of the species, you copied your parents or you died; you ran when your peers ran on the assumption that they knew something you did not; you found where food was by watching where others went. Children learn a huge part of their behavioural repertoire by observing parents, with much of their subsequent personality being based on this learning. Once in school, the child becomes oriented more to the peer group and a vast amount of modelling takes place among children. Television and videos too are major sources of observational learning, and advertisers capitalize on the modelling influence of famous personalities to sell their products. Crazes and fads, such as hula-hoops, skate-boards and roller-skates, are all modelling phenomena, as are the waves of mass hysteria which occasionally sweep through groups of young people. One such event reached the world headlines in 1982 when hundreds of young Palestinian school children fell ill in the Israeli-occupied West Bank Territories. The Israelis

were accused of poisoning the children, but it was later established that this wave of illness was a hysterical phenomenon
spread by modelling, which is more likely to take place in a
highly tense and emotionally charged atmosphere.

As with all types of learning, modelling can be put to good
or bad use and awareness can be present to a greater or lesser
degree. Some modelling is quite conscious, while at other times
it takes place without either person being aware of what is
happening. If you watch two people having a conversation, you
will often notice that they 'mirror' each other's movements and
expressions as they talk. John sits back in his seat, Julie will do
the same; Julie crosses her legs, so does John; John uncreases
his brow, Julie relaxes hers and so on. A boss is less likely to
mirror the non-verbal expressions of a subordinate, but the
underling may well 'shadow' the boss's. Similarly, you are less
likely to imitate the movements of someone you dislike than
someone you are fond of. Try gradually lowering your voice
while talking with a group of friends and you may well find
that, for a short time at least, they will begin to whisper too,
until someone notices and asks why everyone is whispering.

Albert Bandura, of Stanford University in California, is the
main exponent and investigator of learning by modelling. He
reacted against the prevailing climate of American academic
psychology in the 1950s, when it was assumed that humans
learned largely through the direct consequences of their
actions, by demonstrating that people could learn simply by
watching others. This apparent truism does not do justice to
the complexities of the research by Bandura and his colleagues,
in which the parameters and determinants of modelling were
identified and elucidated.[10]

Self-management and self-regulation

In classical conditioning, instrumental learning, and modelling,
change can occur without the full attention of the individual

and at times without any awareness. However, learning of all types takes place more efficiently if attention is focused on the task in question and awareness is present to a greater or lesser degree in most types of learning. Awareness is not an all-or-nothing phenomenon but a continuum which includes a wide range of states of consciousness. Subliminal perception, where one responds to messages one is not conscious of having seen, and sleep-responding, where subjects respond as if they were awake even though asleep, are two examples of the complexities surrounding awareness.

This discussion of awareness is partly meant to dispel some of the myths about behaviourism which imply that behavioural psychologists regard humans as automata devoid of conscious intentions and responsibility for their own fate. Much of human mental and physical activity is involved with the prediction of what is going to happen in the world and with attempts to control events to make survival more likely. This is as true for a baby shaking and sucking a newly discovered toy as for the particle physicist in the laboratory. Rather than trying to monitor and predict every single event, the human brain delegates to semi-automatic processes certain routine functions, the three types of learning already discussed being examples of these. In computer language, they might be termed *sub-routines* and, in the same way as the computer executive function can switch sub-routines on and off, or combine them in different ways with other sub-routines, so we can attend to these semi-automatic processes and manipulate them in similar ways.

The human brain is not a computer, though in some ways it behaves like one. It is an accumulation of successive layers of mental organization, each layer being of increasing complexity and sophistication. In terms of evolution, the last parts of the brain to develop are those involved in the highest forms of human activity—thinking, planning, and anticipating. It is an interesting fact that the last part of the brain to develop in evolutionary terms (phylogenetically) is also the last to develop in the individual human child (ontogenetically). This is in the area of the brain known as the frontal lobes. We know something of what the frontal lobes do by what happens to people who receive injuries there. They become impulsive, distractible, emotionally volatile, and unpredictable; they find it hard to solve new problems and deal with new situations, while well-rehearsed, or semi-automatic, mental routines usually remain relatively intact.

The common feature behind these characteristics is the disruption of high-level monitoring and management of all the constituent functions of the brain. It is as if the general manager of a company is drunk at his desk; the accounts department continues to do its task effeciently, as do the production, sales, and transport departments, but because the boss is drunk, the integration of the various departments is impeded. So long as nothing changes inside or outside the organization, it may roll on, doing a reasonably efficient job; but as soon as something out of the ordinary happens, it will flounder in chaos because the executive part is impaired. The production department piles up goods which the sales department cannot sell, the transport department cannot carry, and the accounts department have not audited.

Thus, the brain is organized hierarchically and our advancement as a species is directly attributable to the development of parts of the brain which can think, not only about the world, but also about ourselves. We can change both, but change is not always as easy as in a computer programme. To extend the metaphor, many of the lower level 'departments' of the brain were once independent agents with no top executive breathing down their necks. They are a bit like successful small businessmen whose firms are taken over by a multi-national company; they may find it difficult to change their ways and often fail to see the rationale for decisions made by the new boss, which they regard as being counter to the interests of their own company. What they cannot see is that these decisions may be in the interests of the larger body.

So it is for habits, or 'semi-automatic functions', whether acquired through classical conditioning, instrumental learning, or modelling. For a number of reasons, they can become as stubborn as the old manager who had worked hard to create the company which is now being changed by the new management. Some habits are remnants of instinctual mechanisms which served the early animals very well. Some forms of anxiety are like this—for instance, the common fear of snakes which occurs in countries where snakes are hardly ever seen. Some psychologists have suggested that certain fears are 'prepared'; that is, for evolutionary reasons unrelated to current living conditions, we are programmed to fear things like snakes.

The development of self-management theory (also termed self-control or self-regulation theory) was partly a response to the obvious fact that behaviour therapists cannot follow their clients around all day, administering reinforcements and presenting stimuli. People had to learn to change their own habits. So why should they not have access to the technology of conditioning in order to gain control over their own errant subroutines? An influential book edited by Thoresen and Mahoney[11] presented a number of empirical and theoretical papers on self-management and, since its publication in 1974, research in this area has mushroomed.

Higher cognitive processes

If the behaviourist movement of the early part of the century was a reaction against the unverifiable and private introspectionism of nineteenth-century psychologists, the cognitive movement of the last two decades is a rebellion against the limitations of a psychology which, in its extremes, regulated human consciousness to the role of an irrelevance. Not that thought was dead in all psychology—psychoanalysts and some social psychologists kept it alive—but it had drifted out of the mainstream of academic psychological inquiry.

As with all good dialectical processes, however, antithesis has resolved into fruitful synthesis, and thought has re-emerged as a respectable topic of scientific inquiry. Some aspects of this resolution have already been touched on in the work of Albert Bandura, who has been one of the leaders in the move away from pure stimulus—response—consequence behaviourism. The work of Michael Mahoney and others on self-control and self-management has pointed in similar directions. Before this, in the 1950s, social psychologists were conducting seminal experiments on the thought processes of human beings in a variety of social situations; attitude formation and attitude change were particularly fruitful areas of study.

One study of thought processes which had a major influence on theory was conducted by Schachter and Singer and published in 1962.[12] This examined how humans interpret what happens inside and outside them, and to what they attribute these events. Volunteers were given injections of adrenalin, a drug which produces non-specific physiological arousal, such as increased blood pressure, heart rate, tremor, and flushing. Subjects were then asked to wait in a room where a 'stooge' was also waiting, and both subjects and stooge were asked to complete a questionnaire. For half the subjects, the stooge acted out a routine of mounting rage at having to fill in the questionnaire, culminating in throwing the paper down and marching out of the room. The other subjects had a stooge who behaved in a euphoric manner and acted in a playfully giddy way. When subjects were interviewed shortly afterwards, those who had been with the angry stooge felt angry, while those with the euphoric stooge felt happy. In other words, given a non-specific state of physiological excitement, the subjects interpreted, or labelled, their sensations according to the cues around them.

This experiment gave additional credence to the view that

people behave intelligibly according to the nature of the mental images they build up of themselves and the world, rather than simply reacting in a reflex manner to stimuli. Interpretations of events are constructed and imposed on the world, as much as they reflect any enduring reality underlying them.

The tendency for humans to see in events what best fits their view of the world is a pervasive one and is an important factor in determining how people behave when drinking alcohol. The picture of the world a person builds up—of attributions and expectations, values and patterns, illusory or otherwise—leads to a more or less coherent course of action guided by this mental plan, subject to the vagaries of established habits and lower-order learned mental and behavioural processes. How efficiently the plan is put into action depends upon the effectiveness of the 'executive processes' (p. 204).

Whether it be the content or the process of thought, conscious mental processes or subconscious ones, they add up to the vastly complex workings of cognition, which determine much of what human beings learn, good or bad. Some more specific principles of cognition and their application to problem drinking will be outlined in Chapter 8.

7 Conditioning

In the previous chapter we gave a brief overview of social learning theory. In this, we shall discuss in greater detail the two earliest types of learning theory advanced by psychologists, classical conditioning and instrumental learning. The deceptive simplicity of these theories is at odds with their importance and complexity when applied to alcohol use.

Classical conditioning

As outlined in Chapter 6, classical conditioning means that, simply by associating two stimuli, one new and one old, the new stimulus comes to evoke the response hitherto produced by the old. How does this apply to problem drinking? All of us who drink will feel like doing so more on some occasions than others. For one person, arriving home after a wearisome day at work may be a 'cue'—a signal or stimulus—for settling down to a pre-dinner drink. Others might enjoy taking a drink with a meal, while those disinclined to socialize might head for the drinks tray at the first opportunity during a party. There are probably as many cues for drinking as there are drinkers. However, some people respond to a greater number of cues than others, and some stimuli are rare while others are pervasive. It is the drinker who has learned to associate drinking with a large number of pervasive cues who will be most strongly and most frequently tempted to drink. It is such people who will find most difficulty in changing their pattern of drinking.

Take Catriona. She is single, 29, works as a personnel manager, and plays a number of sports, including squash and tennis. She likes a few glasses of wine when eating with friends, enjoys a glass or two of lager after playing tennis or squash, and will have a few gins at a party or in a pub with friends. She also meets her friends at home and seldom drinks on these occasions. In her case, drinking has come to be associated with special meals, social occasions, and after-sport relaxation. It happens that she is not an especially gregarious person, plays squash only once a week and eats by herself most week-nights. This means that the cues associated with drinking are not pervasive and, furthermore, are limited in number.

Now take Norman. He works as a solicitor, is single, and lives on his own. It is in the pub that Norman meets his friends and, the way things stand at present, if he did not go there he would not see anyone. He goes to the pub most nights of the week, mainly to meet friends, though, of course, he does drink as well. He also tends to 'have a few' during the not infrequent business lunches to which he is treated by local building-society managers, solicitors, and other professionals with whom he does business. Norman is a gregarious chap who does not like being alone. He forces himself to stay at home one or two evenings a week, but his resolution often fails and he goes to the pub later on in the evening. This is especially true when he is feeling low and he tends to use alcohol as a 'pick-me-up'. He also enjoys a couple of pints of beer most lunchtimes.

What are the cues for drinking in this case? Firstly, all companionship and social meetings are associated with drinking, as are a number of work-related gatherings. For Norman, feeling low and lonely also serves as a cue for taking a drink. These cues are pervasive and numerous and, not surprisingly, he drinks a lot. The fact that he drinks when feeling depressed suggests that, if he goes through a bad patch, he will drink more, and this was indeed the case recently after he broke up with his girlfriend. It is hard for Norman to avoid reminders of alcohol. If he sits alone at home, he feels fed-up, which is a cue for drinking; if he goes to meet his friends, he does so at the local where he meets a barrage of alcohol signals; even at work he is exposed to cues for drinking.

It is certainly not true that the presence of a relevant cue will inevitably make a person drink. It is simply that the desire to consume alcohol is likely to be evoked by cues which have been associated with drinking in the past. Only when an individual sees some good reason to resist that desire, say if a doctor warns that drinking is causing liver damage, is it likely that an attempt to endure the discomfort of resisting will be made.

Almost all drinkers, then, will show some form of conditioned desire to drink under certain circumstances, and the difference between problem and non-problem drinkers in these terms is how numerous and ubiquitous the cues for drinking are. But this is not the only difference. Some drinkers learn, fortuitously or otherwise, that they can relieve the discomfort which arises when their blood-alcohol level is falling. Most drinkers will be familiar with the slight 'seediness' one can experience in the late afternoon after a few lunchtime drinks, while many drinkers will have suffered the torment of a morning hangover. These symptoms arise in part because the body reacts differently to alcohol in the blood when its level is rising from when it is falling, even though the absolute level is the same.

It is true only of some drinkers, however, that these uncomfortable feelings become cues for further drinking. This takes place partly through a process of classical conditioning and partly through instrumental learning. Perhaps fortuitously, say by drinking after work regularly after a lunchtime session, or perhaps deliberately by following dangerous 'hair-of-the-dog' advice, the person learns to associate the discomfort resulting from a waning blood-alcohol concentration with the desire to drink. In this way, a hitherto 'neutral' bodily state is given an emotional tone and acquires drinking-cue status. The reinforcement value of alcohol in its reduction of the unpleasantness of this state also strengthens the link.

In the case of the lunchtime drinker who has an after-work 'curer' and given that the curer is taken consistently at a particular time of day and in particular circumstances, these circumstances may acquire the power to evoke the desire to drink. Figure 7.1 illustrates the development of such a hypothetical link. By a similar process, another drinker might learn to drink in the morning, even in the absence of hangover symptoms.

When cues for drinking are internal, such as sadness, anger, anxiety, or pain, one cannot easily avoid or escape from them, which makes them dangerous and durable cues. Their relative uncontrollability, as compared with cues of place or companionship, means that drinkers are best advised to avoid learning to associate drinking with these internal states. However, there is another important reason why they make dangerous cues. The effects of alcohol withdrawal are very similar to the symptoms of anxiety; shakiness, sweating, tension, stomach upset, dry mouth, nausea, and many other symptoms are common to both.

Generalization

To emphasize the importance of the similarity between withdrawal and anxiety, we must digress a little to describe the Pavlovian principle of generalization. Returning to Pavlov's dogs, let us take the case where a dog has learned to salivate in response to a red light. Through generalization, it may happen that an orange light can produce the same response. Similarly, the dog which has been trained to sit up at the snap of his master's fingers may do so when he claps his hands, since the sound is similar. In other words, cues which are in some way similar to a given conditioned stimulus can become conditioned stimuli simply by virtue of that similarity. This is further illustrated by the hypothetical case of a woman who has learned to drink alcohol to relieve her 'nerves'. Figure 7.2 illustrates what may happen. Once the woman has learned to drink to

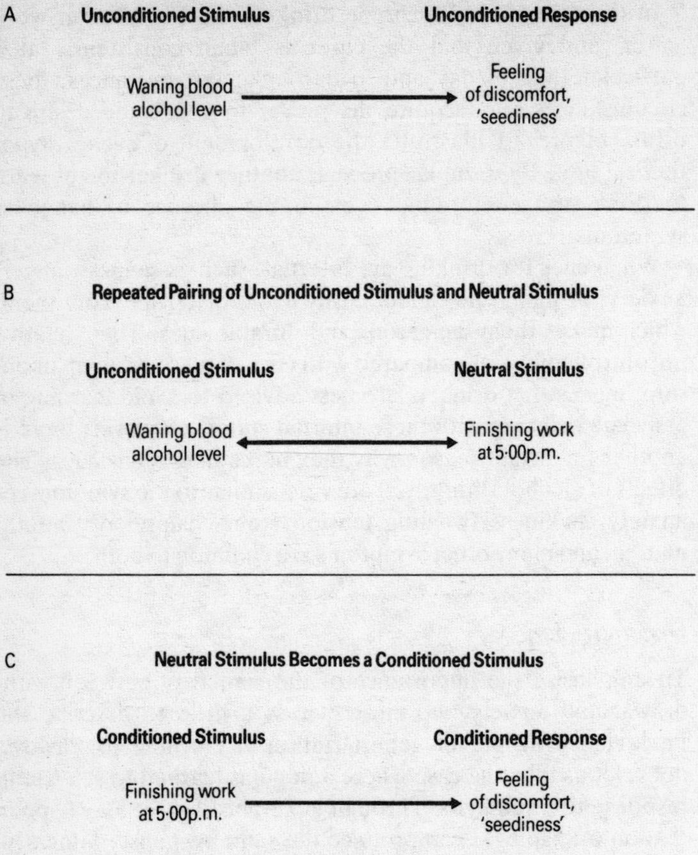

A   **Unconditioned Stimulus**   **Unconditioned Response**

Waning blood ——————————————→ Feeling
alcohol level   of discomfort,
                                                'seediness'

B   **Repeated Pairing of Unconditioned Stimulus and Neutral Stimulus**

Unconditioned Stimulus   **Neutral Stimulus**

Waning blood ←——————————→ Finishing work
alcohol level   at 5·00p.m.

C   **Neutral Stimulus Becomes a Conditioned Stimulus**

Conditioned Stimulus   **Conditioned Response**

Finishing work ——————————————→ Feeling
at 5·00p.m.   of discomfort,
                                             'seediness'

Fig. 7.1.   Classical conditioning of the discomfort associated with a
waning blood-alcohol level.

relieve the symptoms of alcohol withdrawal, a new and dangerous form of learning has taken place. This is the powerful
instrumental learning arising from the rewarding way in which
alcohol reduces its own unpleasant effects. We will expand on
this later.

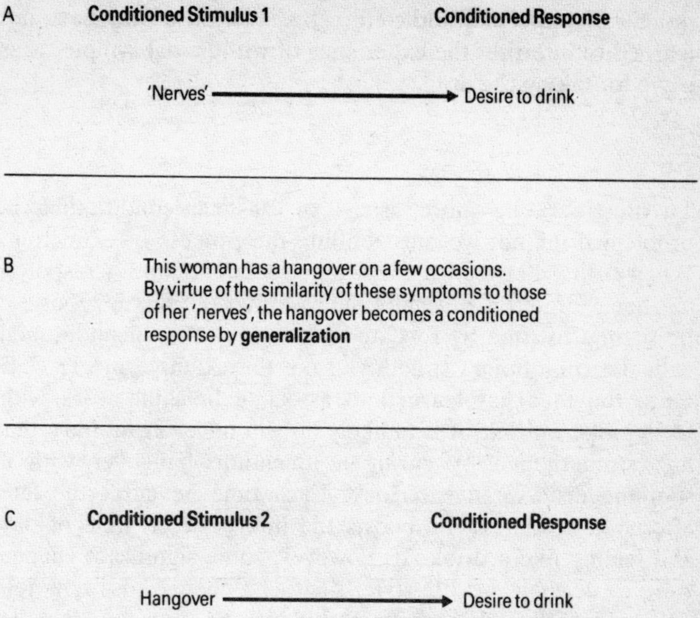

A   **Conditioned Stimulus 1**   **Conditioned Response**

           'Nerves' ————————————————→ Desire to drink

B   This woman has a hangover on a few occasions.
           By virtue of the similarity of these symptoms to those
           of her 'nerves', the hangover becomes a conditioned
           response by **generalization**

C   **Conditioned Stimulus 2**   **Conditioned Response**

           Hangover ————————————————→ Desire to drink

Fig. 7.2. A hypothetical process of generalization.

While these examples may appear simple, they illustrate one important aspect of the new paradigm of problem drinking. Dependence on alcohol is not defined by physiological withdrawal symptoms alone. Rather, problem drinking arises when people learn to interpret these manifestations of withdrawal in a certain way (that is, the symptoms become cues for drinking). Tremor, shakiness, tension, anxiety, sweating, and other symptoms of alcohol withdrawal do not inevitably lead to craving or even a desire for alcohol. Only when the drinker learns to associate them with further drinking do they acquire this power. In the case of opiate-like drugs, patients who are given them for pain-relief in hospital often experience withdrawal symptoms when they are stopped, yet show no inclination to take more—they are not 'addicted'. This is because they have not learned to interpret the experience of withdrawal symptoms as a cue for taking the drug.

Extinction

To illustrate one more aspect of classical conditioning in problem drinking, we must outline the process of extinction. When a link between a conditioned stimulus and a response has been forged, it is usually the case that the link is 'boosted' from time to time by repeated pairings of the unconditioned with the conditioned stimulus. Take the example in Fig. 7.1. Once the man has learned to associate finishing work with feeling like a drink, it is unlikely he will never again have this link strengthened by curing a lunchtime hangover with a soothing drink at five o'clock. Each time he cures the late afternoon seediness, he boosts the link between time of day and feeling like a drink. If, however, some significant change were to occur in his life-style, such as a change to a new job where lunchtime drinking is forbidden, he may find that he ceases to be given unconditioned stimulus 'boosters'. Assuming he does not drink every night, the conditioned stimulus (finishing work) will not inevitably be associated with the discomfort of a waning BAC. In short, the link between time of day and feeling like a drink would be broken. This is directly analogous to the methods by which Pavlov demonstrated extinction. In dogs that had learned to salivate (conditioned response) in response to a bell ringing (conditioned stimulus), when they heard the bell ringing several times without ever smelling any meat powder (unconditioned stimulus), the response to the bell gradually decreased until it stopped completely (extinction). Figure 7.3 illustrates this type of process in the post-work drinker.

There is little doubt that many problem drinkers overcome their difficulties without outside help (see pp. 113–15). More-

A   **Conditioned Stimulus**   **Conditioned Response**

                                           Feeling
            Finishing work ───────────▶ of discomfort,
                                           'seediness'

B   Now the conditioned stimulus occurs several times
            in the absence of the unconditioned stimulus, in this
                    case a waning blood-alcohol level

C   **Conditioned Stimulus**   **Conditioned Response**

            Finishing work ───────────▶ None

Fig. 7.3. A hypothetical extinction procedure: classical conditioning.

over, it is likely that some of this recovery is due to naturally occurring extinction processes. We know that major life-changes, such as new jobs, houses, or spouses, often lead people to change their drinking habits. From a classical-conditioning viewpoint, this makes admirable sense, given that such life-changes often result in the disappearance of certain drinking cues. The newly married woman is no longer alone in the evenings and thus loneliness, which had been a major cue for drinking, is no longer present; the man who has moved house no longer passes a favoured pub on the way home from work; the new job is home-based and the ex-salesman no longer finds himself in commercial hotels of an evening with nothing to do. Simple though these tales are, they are real examples of the types of cue-change which have led drinkers to overcome their problems. Perhaps the best known example is the behaviour of severely dependent problem drinkers admitted to hospital or prison. It is astonishing how uncommon it is for these inmates to complain of a desire for drink which is anything like as great as they would have experienced had they 'dried out' at home. This is largely because they are sheltered from most of the familiar cues for drinking. For this reason, residential treatment of problem drinkers has serious drawbacks, because it tends to build up a false confidence. It is a common misconception among problem drinkers to believe that, because they do not feel the urge to drink while in hospital, they are 'cured'. The resulting over-confidence can mean that they are insufficiently vigilant on leaving hospital and unprepared to cope with the cues and signals for drinking which await them. As a result, they slip quickly back into their old drinking patterns. We will return to the implications of this for the planning of treatment in Chapter 9.

Conditioned tolerance and withdrawal

Until now, we have spoken rather loosely about the possibility of a 'desire' to drink being conditioned to various internal and external cues. It was suggested that alcohol causes certain physiological effects when its level is falling in the blood, and that these effects can themselves become cues for drinking. The severely dependent problem drinker spends much of his or her time attempting to stave off these symptoms by topping up with more alcohol and, as a result, becomes progressively more shaky, anxious, tense, and sweaty. In addition, alcohol progressively loses its power to quell these symptoms. This phenomenon is at the heart of what has been termed 'alcohol dependence', and it has often been assumed that these symptoms are evidence for a physical dependence whose course and nature is largely determined by physiological mechanisms.

Tolerance to alcohol—the progressive reduction in its potency for an individual over time—has been assumed, in particular, to be a good indicator of physical dependence. Recent evidence suggests, however, that these very processes—withdrawal symptoms and tolerance—can themselves be conditioned and learned and are not entirely physically caused. This puts in question the validity of the very corner-stone of the most recent disease models of problem drinking, namely *physical dependence* (see Type C disease concepts, Chapter 3, pp. 79–87). For if tolerance and withdrawal can be learned and unlearned according to psychological principles, so can alcohol dependence.

How can conditioning mechanisms explain tolerance and withdrawal symptoms? Whenever some disturbance is caused in the body, say by a drug, a physiological mechanism known as *homeostasis*, which tries to counteract the effects of the disturbance, comes into play. Not surprisingly, this counteractive process is usually opposite in effect to the action of the drug. So, where an injection of adrenalin causes the production of gastric juices to slow down, the body tries to counter this effect by a compensatory increase in gastric secretions.[1] The net result over time is that the effect of adrenalin on secretions is reduced or even eliminated. Similarly, where a shot of insulin reduces blood sugar levels, with repeated shots the body mobilizes to counteract this effect by producing more sugar.[2] With regard to alcohol, where the effect is to depress the activity of the nervous system, with repeated use the body mobilizes an opposite effect, namely, *activation*, of the nervous system.

The onset of these counteractive processes gradually becomes quicker as the body becomes 'practised'. What eventually seems to happen is that, as soon as the body 'expects' to receive a dose of a drug, it mobilizes its counter-attack when it notices signals or cues associated with past administrations, and this may happen even before the drug is taken. This efficient response obviously has survival value because it prevents the body from being unduly stressed or changed by outside influences.

What happens when the signals appear but the drug does not? The body 'expects' to be given the drug, and so prepares its counteractive response. In the case of alcohol, it activates the nervous system so as to compensate for its anticipated depressant effects. But if no alcohol is actually given, these counteractive processes go unopposed and their effects are not diminished by the pharmacological action of alcohol. The nervous system is in a state of relative excitation, undampened by alcohol. This is what is known as a *withdrawal state*, and the important thing to notice is that withdrawal symptoms are linked to specific cues associated with taking the drug. Such cues might be the taste of whisky, the sound of the pub, and so on. In other words, withdrawal symptoms can be classically conditioned to certain cues. And if they are classically conditioned, then they can be learned and unlearned, generalized, and extinguished, according to the laws of learning already outlined.

According to this model, tolerance and withdrawal symptoms are intimately linked. Tolerance arises because of the homeostatic processes the body produces to counter the action of a drug. These processes produce effects which are, in general, opposite to those of the drug. The growth of tolerance to the drug, that is, the gradual diminution of its effect with repeated use, is caused by the growing strength of the opponent action. The net effect of the drug then becomes its original effect minus the opposing action.

The corollary to this is that tolerance to drugs, and hence dependence on them, is restricted to certain environments— that is, environments where they have previously been taken. There is abundant evidence that animals made dependent on and tolerant to a drug in one environment will lose that tolerance if tested in a different environment.[3]

There is also human research on the experience of heroin addiction by American troops during the Vietnam War which supports this view. According to one major study,[4] roughly one in five American soldiers in Vietnam were addicted to narcotics, the most common being heroin. The United States government was understandably alarmed about this, and preparations were made to deal with an epidemic of heroin addiction when these men returned home. A research study was mounted, and a sample of 439 addicted men, identified as such in Vietnam, was followed up after they had been back in America for a year. The curious fact emerged that only thirty-three were still addicted in the United States, 7.5 per cent of the previously addicted group! This contrasts with the low rates of recovery among men who have become hooked on heroin in the home country. Why should this be? One answer is that the veterans had developed tolerance and dependence in one environment (Vietnam), but on return few of the cues associated with this environment were still present. Hence, they did not experience withdrawal symptoms to anything like the same degree, had they tried to stop while still in Vietnam. It was also notable that 45 per cent of the 439 addicted soldiers used heroin occasionally and without dependence during the twelve months after they left Vietnam. The researchers concluded, 'Contrary to conventional belief, the occasional use of narcotics without becoming addicted appears to be possible even for men who have previously been dependent upon narcotics.'

This conclusion concurs with what we know about alcohol problems,[5] and the lessons from drug research also apply to alcohol. Problem drinkers experience a desire to drink ('craving') when they encounter cues associated with taking alcohol and, over time, they have to drink more and more to achieve an intoxicating effect. If you take them to a totally different environment, however, it is likely that they will show far less tolerance, dependence, and withdrawal, because cues for drinking are absent (as is, in fact, the case when problem drinkers go into hospital or prison).

In short, dependence on alcohol is a learned process and, hence, can be unlearned. One major way of 'unlearning' is through extinction (see Fig. 7.3), or presenting a conditioned stimulus in the absence of the unconditioned stimulus. In this way, the conditioned stimulus loses its power to produce the conditioned response. This suggests that if you repeatedly present a problem drinker with drinking cues, but do not let him drink, these cues will gradually lose their power to produce the excitatory response previously elicited. Later in this chapter (p. 226) we describe a study which shows exactly that.

Let us end this discussion by thinking through the therapeutic implications of conditioned tolerance. If tolerance is indeed conditioned, then extinction should be possible. If extinction can take place, then tolerance can be reduced therapeutically. If tolerance can be 'de-conditioned', so can dependence. For these 'opponent-process' effects of the body to alcohol are nothing else but withdrawal symptoms, the corner-stone of dependence. In other words, withdrawal symptoms from alcohol can be conditioned and extinguished.

The therapeutic implications are obvious. If you present the conditioned stimuli (drinking cues) in the absence of the unconditioned stimulus (alcohol) on many occasions, the conditioned response (excitation, withdrawal symptoms) should extinguish. The 'cue-exposure' treatments which will be described later owe their effectiveness in part to the extinction of conditioned tolerance and withdrawal.

Instrumental learning

At the risk of over-simplification, instrumental learning refers to the way in which the consequences of behaviour influence the likelihood of that behaviour being repeated, in other words, of it being learned (see pp. 199–200). There are four classes of consequence which can affect behaviour, as Fig. 7.4 shows.

	Event is introduced	Event is withheld
Pleasant event	POSITIVE REINFORCEMENT (e.g. smile)	RESPONSE COST (e.g. smile not forthcoming as expected)
Unpleasant event	PUNISHMENT (e.g. criticism)	NEGATIVE REINFORCEMENT (e.g. criticism not forthcoming as expected)

Fig. 7.4. Four types of consequences in instrumental learning.

A consequence can either be pleasant (for example, a smile from someone you respect) or unpleasant (criticism from the same person). It can either occur or fail to occur as expected. The four categories of consequence which emerge from this classification are positive reinforcement, negative reinforcement, punishment, and response cost (see Fig. 7.4). The first two tend to increase the probability of actions they follow, while the other two tend to decrease that probability.

One more principle of instrumental learning should be mentioned before discussing its application to problem drinking. This is *immediacy of reinforcement*. It is firmly established that the sooner a reinforcement follows some behaviour, the more powerful will be its effect on the behaviour and, hence, the more probable that the behaviour will be repeated. The importance of this simple rule for understanding alcohol problems cannot be overemphasized, because the dominance of immediate, short-term reinforcements applies even when longer-term consequences are punishing.

There are at least two ways in which alcohol acts as a reinforcer and, in both of these, its action is sufficiently swift that the immediacy effect can outweigh any detrimental long-term consequence. The first is where alcohol acts as a positive reinforcer, that is, its effect is pleasant and desirable. Many, but not all, drinkers experience the effects of alcohol as pleasant, perhaps not after their first-ever drink but usually after a few 'practice' drinks. People experience alcohol as rewarding in different ways, and a few find its effects punishing or uncomfortable. The initial effect of a small amount of alcohol is to act as a stimulant or 'pick-me-up', but if more is taken the effect reverses and alcohol becomes a depressant. Depressant effects are not necessarily unpleasant, as they can sufficiently dampen down self-awareness for awkward self-consciousness and social inhibitions to disappear.

The second class of reinforcement is called negative reinforcement. By this is meant a consequence which boosts a given behaviour by virtue of the fact that it alleviates unpleasant sensations. When someone drinks to relieve anxiety, the effects may well be negatively reinforcing. Hence the danger that a powerful form of instrumental learning will take place, rendered the more potent because it works quickly to reduce anxiety. It is therefore no coincidence that those who drink largely to reduce anxiety are considerably at risk of developing certain kinds of alcohol problem.[6] This risk is increased many-fold when the unpleasant feelings relieved by drinking are the withdrawal effects of alcohol itself. This is why the morning can of beer taken to quell the shakiness stemming from the previous night's over-indulgence is so dangerous; the act of drinking is powerfully negatively reinforced by the immediate relief provided by the 'curer'.

A fundamental paradox of problem drinking is partly resolved by the principle of immediacy of reinforcement. The paradox is that problem drinkers continue to drink in spite of suffering painful and distressing consequences from their actions. They lose friends, families, and jobs, suffer accidents, diseases, and criminal prosecutions, and become depressed and guilty to the extent of committing suicide in disturbingly large numbers. If the consequences of behaviour govern behaviour, why does this punishment not stop their excessive drinking? The answer is that the punishing consequences come too late. Even if the pain and guilt arrive hours after drinking, this is nowhere near as powerful as those effects which come seconds and minutes after alcohol is absorbed. These effects can take many forms: alleviation of physical pain and discomfort; companionship of the 'old friend'; dulling of worries and guilt over pressing life-difficulties; the warmth and companionship of drinking companions; and many others.

As emphasized in Chapter 6, it is a carry-over from our primitive heritage that these short-term effects often overrule long-term consequences. While it is possible to transcend the immediate in favour of a more worthy, yet distant reward, and while most people learn to do this fairly well as they grow up, it is the burden of the problem drinker that he or she finds it difficult to do so with respect to drinking alcohol. One reason problem drinkers experience this difficulty is that their ability to control thoughts, images, and impulses, and hence control their drinking, is sabotaged by the mental havoc wrought by large quantities of alcohol—not only while alcohol is in the blood but for long periods afterwards. In other words, self-control is inhibited and they find it difficult to engage in the mental activity required to resist temptation and progress towards long-term goals. This does not explain why people become problem drinkers in the first place, but goes some way to explaining how heavy drinking is maintained once started.

Things are not quite as simple as they seem, however, from this account of negative reinforcement. One reason for saying this is that, when severely dependent problem drinkers are observed during the course of heavy drinking sessions, they actually become more tense, anxious, and depressed over time.[7] Some researchers have argued that this goes against the utility of a reinforcement model of problem drinking,[8] but a closer inspection of what happens during a drinking spree does not support this view. Tim Stockwell and his colleagues[9] measured the behaviour of problem drinkers during a drinking bout and found that, while the overall trend was for an increase in unpleasant feelings, the taking of alcohol was associated with a temporary, relative decrease in these feelings following each drink.

This finding illustrates once more how potent immediate reinforcements are. However, there is another factor which contributes to the strength of the tension-reduction, negative-reinforcement process. It has been shown that, in spite of the fact that in the long run most severely dependent problem drinkers feel worse over time, they *expect* to feel better before they drink. Why do they expect something which seldom happens? Part of the answer is that alcohol inhibits memory so that the drinker often forgets how he felt during a bout. Alcoholic blackouts, where the person remembers little or nothing of the time spent drinking, are an extreme example of this phenomenon. We shall return to the role of expectations in problem drinking in the next chapter.

Before discussing the role of the two other types of consequence, punishment and response cost (see p. 235), let us emphasize the continuum between the problem and non-problem drinker. Most drinkers will, at some time, experience a negative reinforcing effect of alcohol—if nowhere else, then simply in cheering up in the pub after a bad day at work. Some drinkers, who could not be sensibly classified as problem drinkers, may even come to depend on alcohol to relieve unpleasant feelings through a process of instrumental learning. Many men and women find they cannot face their mothers-in-law without the fortification of a few drinks. The point is that the quality of the processes involved in drinking in normal drinkers is not different from those applying to problem drinkers. It is the degree to which certain learned patterns of behaviour come into play, together with how these patterns fit in with the whole life-pattern of the individual, which is important. What is happening to the elderly doctor who cannot sleep without his nightcap is not different in kind from what happens to the down-and-out waiting agitatedly for the off-licence to open in the morning to ease his withdrawal symptoms.

Discriminative stimuli

These examples bring us back to the notion of cues or stimuli. Although the process of instrumental learning suggests that consequences are prime mediators of learning, what is learned does not occur randomly. For example, the dog reinforced by chocolate for sitting on command only does so when commanded, but not while crossing the road. It has learned that certain responses in certain situations will produce a reward. These situations act as signals that reinforcement will be coming and are sometimes termed *discriminative stimuli*. It is here that classical conditioning and instrumental learning converge to some extent, as do some of the treatments stemming from the two types of learning process.

As illustrated in the discussion of classical conditioning, cues for drinking are multifarious and idiosyncratic, though among severe problem drinkers they become stereotyped and predictable. Some cues are external, such as the sight and smell of alcohol, while others are internal, such as withdrawal tremor. Many act as discriminative stimuli, that is, signals to the person that reinforcement is forthcoming. The appearance of these cues sets in motion a predictable sequence of actions which results in a reinforcing consequence. So, when the office worker hears colleagues locking their doors at the end of a day's work, not only may these sounds act as conditioned stimuli which provoke a conditioned desire to drink, they also serve as signals that the reinforcing effects of a post-work pint are about to be felt. Less innocently, the experience of waking in the early morning soaked in sweat may serve as a discriminitave stimulus to the severely dependent problem drinker that he is about to benefit from the reinforcing effects of alcohol.

Extinction

Before turning to the therapeutic implications of the instrumental learning account of problem drinking, we must go back to the concept of extinction. It will be recalled that, as applied to classical conditioning, extinction takes place when the conditioned stimulus is repeatedly presented in the absence of the unconditioned stimulus. Similarly, in the language of instrumental learning, a given response will tend to be extinguished when it occurs repeatedly without being reinforced. In evolutionary terms this makes sense, because there is no survival value in persisting to do something which produces no benefits.

The therapeutic implications of this are self-evident. If you expose the drinker to cues for drinking, but discourage him or her from taking alcohol, you have implemented an extinction procedure in which, with time, the cue in question loses its power to evoke the desire to drink. In so doing, you are likely to extinguish both classically conditioned and instrumentally learned responses.

These principles were put to practical use by psychologists Richard Blakey and Roger Baker in Aberdeen.[10] They interviewed a number of problem drinkers to assess which situations would lead them to be tempted to drink. Several situations were listed for each problem drinker and were ranked in order of how 'tempting' the individual would find them. Starting with the least difficult item on their particular hierarchy, say walking down the High Street past familiar pubs, clients were given practice—initially supervised but eventually on their own—at facing up to tempting situations while resisting the urge to drink. Later, some were actually encouraged to go into pubs where alcohol was placed in front of them. It must be noted that each individual had a personal hierarchy of risky situations which was systematically tackled; only when one task had been completely mastered did they turn to the next most difficult. The results of this study were promising. Most clients reported that many of the situations that previously made them want to drink had lost that power to a great extent. It is likely that an extinction process was at least partly responsible for this.

What Blakey and Baker did is not only in accord with reinforcement theory but also follows from the model of classically conditioned tolerance and withdrawal discussed earlier (p. 217). It is probable that they were extinguishing, not only the instrumentally learned response of drinking, but also classically conditioned withdrawal symptoms; hence, tolerance levels may have been reduced as well. Furthermore, they may have helped weaken the link between withdrawal symptoms and the experience of a desire to drink. In other words, the chain of associations linking the various elements of addictive drinking may have been broken in several places.

Hodgson, Rankin and Stockwell have conducted research which goes a considerable way towards supporting the model of alcohol dependence outlined above. Their aim was to weaken the link between internal cues associated with having taken alcohol and the desire to drink. To this end, patients who were severely dependent and who were resident in an alcoholism treatment unit were given fairly large amounts of alcohol (approximately the equivalent of one single measure of spirits for each 7 kg body weight) and then prevented from drinking for the rest of the day. The first time this was done, the men not surprisingly felt much in need of a drink once they had consumed the alcohol. Over the course of six days, however, the desire to drink more alcohol steadily declined. What had happened was that the response (the desire for alcohol) to the internal, physiological sensations of having taken alcohol had been extinguished to a considerable extent. Figure 7.5 illustrates this cue-exposure process. It should be noted that, after

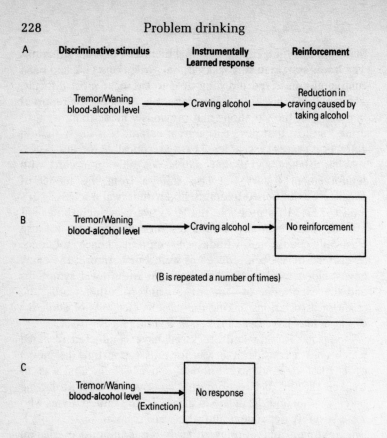

Fig. 7.5. A hypothetical extinction procedure: instrumental learning.

drinking their priming dose, patients were asked to hold, look at, and smell another glass of alcohol but not to drink it. Thus, these men were also being exposed to external cues associated with drinking.

This research shows that the 'loss-of-control' phenomenon, hitherto thought to be a result of a purely biologically induced craving, is in fact a learned process which can be unlearned. If the theoretical model of tolerance and withdrawal (p. 216) is valid, then some extinction of tolerance and withdrawal would also be taking place on the occasions when clients had to smell, hold, and look at a drink without drinking it.[11]

Reinforcements and social stability

If the reinforcing properties of alcohol lay only in its immediate, psycho-physiological effects, drinking problems would be far less troublesome than they are. Support for this assertion comes from the fact that most problem drinkers do, at times, stop drinking for a while, if for no other reason than that they are feeling physically ill or have been admitted to hospital or prison. Many, if not most, 'relapses' into heavy drinking following abstinence occur in the absence of any physiological disturbance caused by withdrawal. This happens partly because of an abundance of environmental cues and discriminative stimuli but also because the use of alcohol is reinforced in many other ways.

It would be stating the obvious to describe how, particularly among certain sub-groups of society, people are expected to drink alcohol and how its use is socially reinforced. This is as true of young, male, manual labourers as of Members of Parliament. What is less obvious is how little reinforcement some people have for doing things other than drinking. It is an unfortunate fact that, in many socially impoverished environments, there are few social activities which even approach drinking in their rewarding value—and this is particularly true of problem drinkers who have gradually lost interest in hobbies and sports, alienated their friends, and separated from their families. In so doing, they have given up a large proportion of the natural rewards which keep most people on the 'straight and narrow'. Readers may care to ask themselves what *they* would do in such a position. Alcohol and the places it is drunk are so accessible and ubiquitous that the 'reinforcer less' individuals of our society easily take recourse to the solace and cheer they provide, however short-lived these may be.

When someone is thought to be drinking too much, one of the most useful questions which can be posed is, what advantages are there for this person in cutting down or stopping? If the answer is 'very few', not much in the way of useful counselling can be offered which is not directed specifically at finding reinforcers for a changed pattern of consumption. Too often, clients who can see little reason to change have been exposed to a programme of moral exhortation in which they are condemned as unmotivated.

It is no coincidence that 'social stability', which is virtually synonymous with having access to a wide range of society's reinforcers, is one of the best predictors of recovery from alcohol problems. Where there is a reasonably intact marriage, a satisfying job available with no financial problems, and access to a group of supportive friends and social activities, recovery from alcohol problems is, in many cases, not too difficult. For most severely dependent problem drinkers, the physiological effects of alcohol are the least of their difficulties. It is access to reinforcers for abstinence or moderate drinking, whether material, social, or psychological, which is the biggest stumbling block. Those experts who equate alcohol problems with a narrow range of physiological symptoms have done a great disservice to problem drinkers by obscuring this fact.

Sometimes, however, reinforcers for drinking are less obvious, and this was demonstrated in a study by Orford and Edwards[12] of 100 married, male, problem drinkers. Using a method called the 'semantic differential', each wife and husband was asked to make independent ratings of how the husband behaved in three conditions: when he was drunk, when he was sober, and how ideally they would like him to be. Two aspects of these ratings were of particular interest—how affectionate and how dominant (or assertive) he was seen to be in these three conditions. Not surprisingly, wives tended to rate their husbands as being further away from the ideal in affection when drunk than when sober. What was more interesting was that the women saw their partners as being nearer the ideal in dominance when the men were drunk than when sober. This was how the men saw it too, according to the ratings they made about their own behaviour. In short, as far as assertion and dominance were concerned, both partners thought the man improved when drunk.

This is an interesting example of a non-obvious positive consequence for drinking and a powerfully immediate one at that! Within at least some of these families, the man would behave more assertively when drunk than sober, and thus the family system was structured in such a way as to reinforce heavy drinking in the husband. It might therefore be possible to improve the success rates of treatment if the men to whom this applies could find some means of being assertive other than by becoming drunk. One possibility is *assertion training*, a variant of social-skills training in which clients are given practical instruction in how to improve their communication and assertion techniques.[13] Something similar to this, known as *communication skills training*, can be carried out in a family context.[14]

The practical implications of an instrumental-learning model of problem drinking were elegantly, if expensively, implemented in the United States by two clinical psychologists, Nathan Azrin and the late George Hunt.[15] They selected eight hospitalized problem drinkers with severe problems and put them into what was termed a 'Community-Reinforcement Programme'. The overall aim was to ensure that the men were predictably and consistently reinforced for abstaining and that these reinforcements were not available when they were drinking. This is in marked contrast to what happens in the natural world of most problem drinkers who often receive more attention from medical and social services, and even families, when they are in 'crisis' and drinking heavily than when functioning well and not drinking. Similarly, spouses of problem drinkers oscillate, understandably enough, between punitive and caring responses to their partner's heavy drinking.

Hunt and Azrin sought to eliminate this sort of inconsistency by acting as 'social engineers' and virtually reconstructing a social niche for each of their clients. First, the men were given vocational counselling, helped to find work, and given inter- view- and job-seeking skills training. Second, intensive marital and family counselling was carried out with those who had families still willing to receive them at home. If a man had no family, a 'foster' family was arranged, consisting of people who might have some reason for keeping in contact, like relatives, ministers or priests, and even employers. These foster families were encouraged to invite the client to dinner on a regular basis and to expect him to help in chores and offer services in other ways.

In all these families, foster or real, the strict advice given to family members was the same: for as long as the client was drinking, interaction was to be reduced to a minimum or even stopped; when drinking ceased, arrangements were to be resumed. In the case of wives, this could mean leaving home to stay with relatives or in a motel.

To reconstruct a social life for the men, a 'dry club' was set up where they could meet, play cards, listen to the juke-box, invite friends, have picnics and outings and enjoy a host of other social activities. This was a self-supporting organization to which clients had to pay monthly subscriptions after the first month's free membership. Any member who showed signs of drinking was turned away. In addition, counsellors helped clients, even financially, to obtain cars, telephones, newspaper subscriptions, and other necessities of the American way of life. They were visited twice a week by a counsellor during the first month of discharge and subsequently twice per month on average.

The results were commensurate with the expense and effort involved. Compared to a matched control group, the target client spent six times less of their time drinking during the next

six months and differences between the groups on other measures were correspondingly large. There can be little doubt that
an expenditure of resources on this level as a routine measure
is simply not economically or politically feasible. Nevertheless,
it is possible to envisage programmes at the social and individual level, based on the same principles as in the Hunt and
Azrin study, which confine themselves to existing resources.
For instance, if the client has a spouse, a contract can be drawn
up between the husband and wife making it explicit that the
spouse will withdraw as soon as the client begins drinking, but
cease from recriminations about past behaviour when he or she
is not. As well as establishing consistency of reinforcement, this
has the added effect of reducing some of the spouse's anxieties
by giving guidelines for how to respond.

When problem drinkers have alienated friends and relatives
by past behaviour, it may be possible to re-establish some of
these relationships through joint meetings with the counsellor,
client, and the said friends or relatives, if any are willing to
attend. Contracts similar to the one described for husband and
wife may be set up. If friends know they do not have to tolerate
the drunken behaviour of the client and if they no longer feel
guilty about turning him away when drunk, they may be more
inclined to offer help and support when he is behaving in an
acceptable way.

Contracts not dissimilar to this have been used for some time
in organizations operating alcoholism policies for their employees. Workers who are persistently late, absent, or inefficient
because of drinking, or who are affected by alcohol during
working hours, are offered counselling as an alternative to
normal disciplinary action. In the background, there is always
the sanction that, if the client persists in drinking to a degree
that work is impaired, disciplinary action will be instituted.

There are many other ways in which 'community reinforcement' can be implemented in a normal counselling environment
without the expenditure of extra resources. At a social level,

however, a fundamental re-orientation of the approach to alcohol problems could be envisaged on the basis of this model. In particular, the tendency for clients to receive help and attention from agencies only when they are drinking and failing to cope reinforces maladaptive ways of behaving. With many agencies, once clients are discharged, they will be ignored to the extent that they succeed in coping with their problems. If they have families and friends who reward coping behaviour, there is no great harm in this. But if they are socially isolated, as many problem drinkers are, they may relapse and have this failure reinforced by the renewed interest of counselling agencies. If some way could be found, perhaps through an intensive mobilization of voluntary, community, and professional resources, to build up a system of predictable and consistent community reinforcements, contingent on coping and the avoidance of harmful drinking, many relapses might be prevented. However, this is speculation and must be evaluated in a scientific manner.

Spontaneous remission

It is not only in the development and maintenance of problem drinking that instrumental learning is important; it is also crucial for the way those giving up harmful drinking patterns learn to overcome their problems. It is likely that as many people give up problem drinking on their own every day as do so with outside help, if not more. This phenomenon of 'spontaneous remission' (see pp. 113–15) often occurs when people change jobs, establish a new loving relationship, change where they live, or simply begin to 'see things in a different light'. In one study where recovered clients in a treatment programme were asked retrospectively what had helped them get better,[12] hospital treatment, counselling, and AA all came at the bottom of the list. At the top were the sort of largely external changes just mentioned.

How this fits in with instrumental-learning theory is obvious. People will change their drinking patterns if they see some benefits from doing so, that is, if they anticipate receiving reinforcement for a new life-style. Meeting a person with whom one can form a caring and loving relationship is one of the most common ways in which people learn that another life-style is possible, where pleasure comes from something other than drinking. Finding a satisfying job can have much the same effect, as can any number of other significant life changes.

The third type of instrumental-learning process (see Fig. 7.4) must be introduced at this point. The meaning of punishment is self-evident—the application of an unpleasant event which reduces the likelihood of repetition of the act it follows. Punishment, like positive and negative reinforcement, is an important determinant of whether or not problem drinkers change their drinking habits. One common form of punishment is physical ill-health and the associated pain, sometimes acute and immediate, which comes with drinking alcohol; not a few problem drinkers report that a deterioration in their health caused them to give up or cut down. In addition, some drinkers do respond to punishing criticism and withdrawal of love and friendship from family and friends. If that withdrawal is consistent, immediate, yet selective (withdrawal which makes it clear that it is the *drinking* that is unacceptable, not the *drinker*), it may overcome the powerful positively and negatively reinforcing effects of alcohol. This withdrawal of pleasant consequences is termed response-cost, the fourth type of consequence outlined in Fig. 7.4.

In this light, it can be seen that kindness which takes the form of collusion with drinkers in their drinking, together with protection from the consequences of their behaviour, can in the long run be a cruelty, as it may impede natural learning processes. Sad to say, some treatment programmes have a similarly counter-therapeutic effect by cushioning the drinker from the effects of drinking.[16] We will discuss the implications of this for policy-making in Chapter 9.

The circumstances under which punishment and response-cost overwhelm the immediate positive and negative reinforcement power of alcohol depend on a number of factors. When punishment is immediate, say an acute stomach pain whenever alcohol is taken, this may well take precedence over the reinforcing effects. Even when punishment is less immediate, it may eventually be sufficiently unpleasant that it finally 'gets through' to the drinker. Stories of how problem drinkers reach 'rock bottom' and realize that they must change are familiar and, while such stories have been oversold in being applied to *all* problem drinkers, they are true for some. They provide a good example of how the balance of reinforcement can change to produce a change in behaviour. Of course, many other things are happening when such 'conversions' take place and, in the following chapter, we turn to some of the higher forms of learning which are involved.

8 Cognition

Having covered the two earliest forms in the last chapter, we now ascend the phylogenetic scale and turn to higher forms of learning. The relationship between the various forms of learning is complex and inadequately understood. What we can be certain of is that there is no contradiction in using both conditioning and higher forms of learning to explain alcohol problems; indeed, we need both to come anywhere near a proper understanding.

'Cognition' is an umbrella term to describe the higher mental processes. We do not use 'thought' because it implies the presence of conscious awareness. We are, in fact, unaware of much of our mental activity, and thus 'cognition' is a less misleading term than 'thought' for the variety of mental processes covered in this chapter. We begin with modelling, then deal in turn with self-regulation and the higher cognitive processes.

Modelling

When alcohol is advertised, attractive or high-status people are usually seen to be rewarded in some way for drinking it, whether through manly strength, feminine allure, social accomplishment, or the prospect of sexual adventure. Depending on the type of drink and the target group aimed for, certain qualities are attributed to those who use the brand in question and these qualities are sure to be seen as desirable by most of the target population. In using these methods, advertisers are tapping a primitive human instinct—the tendency to model one's behaviour on successful and respected contemporaries.

This propensity has had in the past, and to some extent still has now, immense survival value for the human species, with much of what we learn being acquired through observing other people. This is no less true of drinking habits than anything else. The fact that drinking patterns in general, and drinking problems in particular, tend to run in families bears testimony to this point. While there can be little doubt that there is some genetic loading in the transmission of drinking problems within families, it is unlikely that this accounts for more than a small proportion of acquired drinking patterns (see Chapter 4).

One aspect of modelling is the simple acquisition of a style of drinking by children observing their parents, relatives, and older siblings. To give two elementary examples, if a father habitually drinks ten pints of beer on a weekend evening, it is possible that his son will follow suit; if a mother drinks alone at home, her daughter may behave similarly at a later date. The use to which alcohol is put by parents may also be modelled. Where parents react to life problems by getting drunk, or where anxiety and depression are 'self-treated' by generous prescriptions of alcohol, children may pick up this style of coping, and it is this which may be transmitted rather than any particular quantity or pattern of drinking.

Another way modelling may be involved is a negative one, relating in this instance to a *failure* of modelling. By this, we mean a failure to acquire certain psychological qualities which protect the observer from the tendency to drink too much. It is well established that many children of serious problem drinkers encounter multiple difficulties at home and school; they tend to be isolated, badly behaved, and to be under-achievers at school, although these ill-effects may only manifest themselves when there is family conflict and aggression associated with drinking. It is also well established that children who are badly behaved, delinquent, and under-achieving are most at risk for developing alcohol problems later in life.[1]

There may be a common factor linking these diverse behaviours. Perhaps one of the reasons children of problem drinkers tend to show behaviour disorders and to develop alcohol problems in later life is that, owing to the disorganization of their family lives, they fail to acquire a solid set of internalized standards by which to monitor and judge their conduct. If one parent is deceiving the other, say about drink or debts, if a parent makes promises about stopping drinking which last only until the following Friday, or if physical violence is frequently taking place, it must be difficult for children to acquire a set of standards and guidelines for their behaviour. If these standards are poorly developed, it is likely that children will run into trouble with their peers, the law, employers, or other members of society and will become more liable to prosecution, divorce, unemployment, or friendlessness. It is precisely this kind of person who is detached to some extent from the framework of society and, consequently, deprived of many of society's rewards. For all the reasons mentioned in the previous chapter, such a person is abnormally vulnerable to the easy and accessible reinforcement of alcohol and its drinking culture. Perhaps it is also partly for these reasons that the associations between problem drinking, crime, divorce, and unemployment are so strong.

It would be misleading to suggest that this type of 'negative modelling' mechanism is the principal way alcohol problems arise. If it were, how would we explain the fact that certain occupational groups, such as doctors, journalists, and publicans, are disproportionately at risk for developing alcohol problems? It is very unlikely that these groups have abnormally disturbed family backgrounds and their elevated risk of problems is more readily explained by the increased availability of alcohol and other social factors.

Among the factors which predispose some occupational

groups to drinking problems is that they model the behaviour of their professional peers and seniors. In the case of journalists, drinking seems to be an important medium for news gathering and cub reporters learn to follow in the steps of their more experienced colleagues. In the armed forces, a similar 'drinking culture' exists where, not only is passive modelling of heavy drinking a common occurrence, but there is also active pressure on the new recruit to conform to established heavy-drinking norms.

Peer groups of all ages are important determinants of drinking patterns, and attitudes to drinking may be transmitted in addition to actual drinking habits. Favourable attitudes to alcohol are a crucial predictor of heavy drinking and are acquired largely through modelling.

It is not only those who are important to a person, however, who can influence his or her drinking. One study carried out in the United States showed that, under certain specified conditions, normal people can be influenced to drink at near alcoholic levels if in the presence of those who are drinking heavily.[2] If modelling can be as powerful as this between total strangers, its effects among families and groups of close friends and acquaintances must be considerable. Bandura's work (see Chapter 6) shows that modelling will occur more readily if the model is of high status and liked by the observer or if the model is seen to be rewarded for the behaviour in question. It is not difficult to grasp how these conditions apply to those whose drinking is moulded by their families and peers or to appreciate the power of the learning which takes place in these intimate contexts.

An interesting example of the transmission of drinking practices is given by McAndrew and Edgerton in their book *Drunken comportment*.[3] They cite historical sources to show how some tribes of American Indians changed the way they used alcohol as a result of observing how white men drank. Apparently, many tribes already used naturally fermented alcohol for ceremonial purposes before the arrival of the Europeans, but aggressive drunkenness was virtually unknown. But by modelling their behaviour on that of the powerful invaders, they quickly learned what the effects of alcohol 'should be' and took to the new behaviour pattern with frequently disastrous results.

This example demonstrates, not only that drinking habits can be transmitted by modelling, but also that the kind of behaviour exhibited when drunk can be learned in this way. In many Western societies, some learn to be aggressive and uninhibited when drunk, while in other cultures, as McAndrew and Edgerton have shown, exactly the same chemical evokes passivity and introspection when taken in large quantities.

Self-regulation

While there are many problem drinkers who fail to acknowledge they have a problem, it is also common for others to complain that they simply cannot stop drinking heavily, despite wanting to badly. These people are reporting a failure of their self-regulatory system, whereby semi-automatic, habitual behaviour patterns have come to function outside conscious control. There are a number of components to the self-regulatory, or self-management, system, which will be discussed. These are self-monitoring, self-evaluation, stimulus control, and self-reinforcement.

Self-monitoring

Self-monitoring is the process whereby we pay attention to what we are doing, so that we become aware of when, where, and to what extent we are doing it. We cannot possibly be aware of everything we do at every moment, for that would overload the brain. So it is mainly when we have to make changes in habitual behaviour patterns that we must begin thinking about what we are doing and how to change it. Of course, periodic self-monitoring to check that all systems are functioning well is desirable, but in many cases it does not happen. This failure in self-monitoring can be a major cause of people slipping into 'bad habits', such as smoking, drinking heavily, nail-biting, and many others.

Therefore, one important purpose of self-monitoring is to act as a kind of 'safety-check', to let people know whether whatever it is they are doing is staying within the bounds of what is acceptable to themselves and others. (Such a safety-check presupposes that they know what is safe or desirable and what is not; this is a question of self-evaluation which will be discussed shortly.) In respect to problem drinking, this function of self-monitoring refers to knowing how much one is drinking. Among heavy drinkers, underestimation of alcohol consumption is common, and this failure of self-scrutiny is probably a major cause of drinking problems. Conversely, encouraging self-scrutiny of this kind is of potentially great importance for preventing more serious problems from arising.

A second purpose of self-monitoring is to make oneself aware of the circumstances under which a given habit tends to occur. Take the example of binge eating. Some people are prone to engaging in eating 'orgies' without being aware of what triggered off the binge. Boredom may be the trigger for one person and indecisiveness for another, though the exact nature of the trigger will only become apparent once the individual had carefully monitored his behaviour. Similarly, drinkers may be unaware of the particular cues which make them feel like taking alcohol. Once a person knows where, when, and under what circumstances a habit is triggered, these conditions may be altered as a means of changing the habit. Certainly, many drinkers *are* well aware of what makes them want to drink, but others are less enlightened and formal self-monitoring exercises can be of great value.

Not all cues stimulate the urge to drink excessively. There are moderate drinking cues and heavy drinking cues, though their nature will vary considerably from person to person. Fig. 8.1 gives an example of a self-monitoring exercise for one heavy drinker who was attempting to identify circumstances associated with heavy as opposed to moderate drinking. The exercise required him to fill in details of the last four occasions when his drinking caused problems and the last four when his use of alcohol was moderate and trouble-free.

It is clear from Figure 8.1 that the cues associated with heavy drinking for this man were as follows: (a) starting drinking before 7 p.m.; (b) drinking with Colin; and (c) engaging in 'pure' drinking, that is, not doing anything else at the same time, such as playing darts. Obvious as these may be, the client in question was not aware that he tended to drink heavily in these circumstances until he carried out the exercise. He was able to use this information to plan to avoid these cues.

Self-evaluation

This is the next stage in self-regulation and refers to the process of comparing one's performance with a predetermined, internalized standard or goal. In other words, it means judging how one has done in the context of how one thinks one should have done. Once an alcohol problem has been established an erosion of personal standards may well take place, even in those individuals whose standards were intact and well-developed before the onset of the problem. This is an understandable consequence of the fact that drinking can become one of the highest-priority activities for the drinker, taking precedence over family, friends, work, self-respect, health, and leisure pursuits. In the same way as starving people will often abandon the luxury of moral standards and self-respect to find food, so severely dependent problem drinkers may have recourse to deceit and stealing to secure a supply of alcohol. In their eyes,

FOUR TIMES WHEN MY DRINKING WAS TROUBLE FREE

	Day	Date	Time	No. of hours drinking	Place	People	Other activities	No. of units alcohol	Money spent on alcohol	Consequences of drinking (if any)
Time 1	Thu.	2/4	8 –11p.m.	3	Social Club	Darts team	Darts	8	£3·00	
Time 2	Fri.	10/4	8 –12p.m.	4	Chinese Rest'rant, Kingfisher bar	Wife, Betty, Bill	Eating	11	£4·50	
Time 3	Sat.	14/5	9 –12p.m.	3	Social Club	Bill, Archie	Dominoes	10	£5·25	
Time 4	Sun.	15/5	8 –11p.m.	3	Kate's bar	Alister, John	None	8	£3·50	

FOUR TIMES WHEN MY DRINKING CAUSED ME TROUBLE

	Day	Date	Time	No. of hours drinking	Place	People	Other activities	No. of units alcohol	Money spent on alcohol	Consequences of drinking (type of trouble)
Time 1	Fri.	3/4	6 –11p.m.	5	Archie's flat, the Horseshoe	Archie, Bill, Colin	None	20	£11·00	Fell, cut my hand
Time 2	Sat.	11/4	5 –10p.m.	5	Kate's bar, Chinese Rest'rant	Bill, Colin	Fruit Machine	16	£9·50	Argument with Bill – went home early
Time 3	Thu.	16/4	7 –11p.m.	4	Social Club	Alan, Jim, Colin	None	16	£10·25	Hangover – late for work next day
Time 4	Sat.	21/5	6 –12p.m.	6	Social Club	Colin, John, Alister	None	22	£14·75	Missed bus home – argument with wife

Fig. 8.1. Self-monitoring exercise for a heavy drinker.

drink is as vital as food to the starving. Similarly, the drinker may cease to place much importance on achieving in other areas, say in sport or even as a parent, because the desire for alcohol supersedes these bases for self-evaluation.

Once drinkers are in this position, not only are they 'reinforcerless' but they are also in something of a personal vacuum, deprived of some or all of the personal standards, moral or otherwise, by which to judge themselves and hence gain self-respect. It is not surprising that Orford and Edwards (see p. 230) found that the self-esteem of their problem drinkers was a significant predictor of a successful outcome. The woman whose husband has left her because of her drinking, whose children are living with their grandparents, and who has alienated her friends and lost her job, has precious few sources of self-esteem. And where self-esteem is low, the person will see little point in trying to improve things and overcome her problems, and will disbelieve in her capacity to do so.

It follows that one of the main aims in counselling problem drinkers should be to help them set achievable goals, through the realization of which they can boost their self-esteem. Such goals may be related to family, leisure, work, finances, or drinking itself. Meaningful targets are often difficult to identify, however, because problem drinkers frequently have problems in some or all these areas. It is, for example, common for problem drinkers gradually to give up sports and hobbies as their drinking increases. They often get so unused to taking part that, when they stop or reduce drinking, they despair of being able to occupy their time in any other way. The resulting feelings can further weaken self-esteem and increase the difficulty of finding routes for achievement.

Later in this chapter we shall describe the phenomenon of 'learned helplessness'. This is a psychological state whereby a person comes to believe that nothing he does will influence what will happen to him. Many problem drinkers find themselves in this position; the obstacles they see ahead of them are so large that they do not even make an attempt to overcome them. One way of helping people break out of this trap is to set intermediate, short-term goals, which are not seen as insurmountable precisely because they are so modest. This kind of goal-setting is a central feature of a good self-regulation, and is something most of us do much of the time in various ways. We break up tasks into sub-tasks and reward ourselves with a rest or some distraction when one of them is completed, whether this be while digging the garden, painting a wall, writing a long report, or going for a hike. If we fail to set ourselves sub-goals, we can easily become demoralized by the apparent inaccessibility of the ultimate goal.

Some people are better than others at this aspect of self-regulation, and those less accomplished may need outside help in setting achievable sub-goals. One study[4] showed how helpful this approach can be with a group of overweight individuals who were trying to cut down eating. Clients were divided randomly into two groups, and those in the first were asked to count how many mouthfuls of food they ate each day for four weeks; their task was to reduce by 10 per cent the total number of mouthfuls taken each week for the four-week period. The second group made similar recordings, but were asked to reduce by 10 per cent the number of mouthfuls taken during each four-hour period of every day. The first group did not lose weight but the second did, largely because they received frequent and immediate feedback about how they were faring, thus boosting their resolve, morale, and sense of achievement. The other group had to wait for a week before making the calculations which would tell them whether or not their performance was matching up to the goals they had set themselves.

How does this apply to problem drinking? In the first place, as Ray Hodgson has pointed out,[5] the method used by the successful group in this experiment is almost the direct equivalent of the AA principle of 'one day at a time', and the usefulness of this mental strategy is acknowledged by many AA members. In controlled-drinking programmes, goals are set for drinking levels which are highly specific;[6] daily, and even hourly, limits are set by the clients, so that they get immediate feedback about their performance. Short-term goals can be set equally well for activities other than drinking. To help a drinker re-acquire the taste for engaging in some former sport or hobby, for example, a number of short-term goals may be set which take them closer to full reinvolvement.

The cue-exposure treatments described in Chapter 7 may partly owe their effectiveness to the cognitive effects of immediate feedback for achieving short-term goals, namely, resisting the urge to drink in the face of drinking cues. From breaking up major tasks into small sub-tasks, drinkers may gradually have gained a sense of mastery over their impulses and the resulting confidence may be as important as any extinction processes occurring simultaneously.

Stimulus control

The concept of stimulus control refers to the situation in which individuals can themselves change their behaviour by altering stimuli in their environment and, as a consequence, change classically conditioned or instrumentally learned responses. In other words, stimulus control means arranging the cues in your environment so that you are more likely to do what you would like to do and less likely to do what you would rather not. In the case of learning how to study efficiently, for instance, it helps to have a particular desk or table set aside for studying alone and for nothing else; in so doing, cues arising from sitting at the desk become associated with studying—rather than eating, day-dreaming, watching television, or other activities incompatible with studying. Similarly, overeaters are encouraged to restrict eating to a particular table or chair and to avoid eating in locations where other things are usually done. By confining eating to a limited number of locations, the habit becomes linked to a restricted set of cues, and commonplace situations like watching television or sitting in an armchair gradually cease to be connected with the consumption of food. The application of stimulus-control procedures like these to problem drinking has been promising and can be used both in abstinence and controlled-drinking programmes. To take the example given in Fig. 8.1, a simple stimulus-control regime would be to encourage the client to avoid drinking with his friend Colin, not start drinking before 7 p.m. and give up drinking when that is the only thing he is likely to be doing.

Self-reinforcement

The final stage in the self-regulation process is self-reinforcement, that is, the rewards and punishments we administer to ourselves in response to how we think we have performed. These reinforcements need not be material and are, in fact, largely intangible or symbolic. We praise ourselves silently with complimentary words and images, or we castigate ourselves with reprimands and negative thoughts and images. Intimately connected with self-evaluation though self-reinforcement obviously is, there is one important distinction. This is that it is possible to evaluate what one has done as satisfactory, yet refrain from giving oneself a reward—whether that be a rest, a drink, a smoke, or any other of the myriad pleasures we use to reinforce ourselves. Alternatively, in spite of acknowledging that he has failed to meet a particular standard, a person can indulge himself with a reward of some kind; 'spoiled' children, for instance, learn to do this all too well. In a similar way, individuals can punish themselves more or less generously.

Self-reinforcement can act both as a motivator and as a signal that one has achieved a particular goal. It can therefore act in its primitive 'energizing' fashion, similar to the way externally imposed reinforcement works, as well as serving the function of highlighting the achievement of a particular sub-goal. While self-reinforcement is largely symbolic, and though it is so intertwined with self-monitoring and self-evaluation as to be indistinguishable from them at times, it can be useful on occasions to introduce 'artificial', tangible self-reinforcement. This is particularly appropriate when some entrenched habit is resisting the blandishments of the person's conscious attention and desire to change. Self-reinforcement procedures have been shown to enhance significantly the effectiveness of a number of self-management programmes, particularly in weight-reduction schemes,[7] though its application to alcohol problems has yet to be properly tested.

The same guidelines for effective reinforcement apply when it is self-imposed as when externally administered; in particular, the interval between the behaviour and the response is very important. In designing a self-reinforcement programme for, say, someone who wishes to stop drinking completely, the counsellor should try to devise reinforcements which are readily accessible and which do not require much effort to obtain them. The client may be encouraged to break down the day into three periods—morning, afternoon, and evening—and make it a goal to abstain for the next five hours. (This is again a refinement of the AA advice, 'One day at a time'.) At the end of each period, the person would be encouraged to reward himself or herself in some way—perhaps by putting a fixed amount of money in a jar which was to be spent, at the end of the week or month, on some luxury. Alternatively, rewards could be delicious foods, non-alcoholic drinks, or an outing to the cinema, art gallery, sauna, snooker hall, and so on.

While self-regulation is a fundamental part of everyone's day-to-day life, many problem drinkers find themselves with so many troubles that they lose the incentive to mesh their behaviour with what is required by personal and social standards. It is quite common for a disabling apathy to set in, with its associated low morale and low self-esteem. For problem drinkers to put self-regulation activities into action, they must see some reason for doing so; they must expect some change in their lives—whether in terms of relationships, employment, money, housing, leisure, health, or all these things. Unless improvements can be shown in at least some of these areas, self-management programmes are doomed to failure. In the case of problem drinkers who do have such incentives to change, self-regulation programmes can help them uproot old habits and sow new ones.

Higher cognitive processes

Problem drinking is not common among sub-human animal species. While such species obviously do not have the ability to produce alcohol, fermented fruit with a high alcoholic content is available naturally in many parts of the world and many animals could drink it if sufficiently motivated to do so. It is also the case that it is remarkably difficult to get animals voluntarily to consume alcohol in laboratory conditions. This suggests that the things which best distinguish us from the sub-human species—our thought, language, and intelligence—are somehow related to why we drink alcohol and why some of us drink so much of it.

Self-awareness

One faculty which is a function of our intelligence and which further distinguishes us from animals is a highly developed self-awareness. Has this anything to do with our liking for alcohol? One author[8] has amassed a great deal of evidence to suggest that a major reason why people drink is to reduce self-awareness, and thereby escape many of the stresses, worries, and anxieties which inevitably arise from it. In one study, problem drinkers were followed up over three months after treatment. They were divided into two groups, one with low and the other with high self-awareness, as measured by a questionnaire assessing the extent to which an individual habitually engages in self-scrutiny. Those subjects in the high self-awareness group who had also experienced a large number of stresses since discharge showed a high relapse rate of roughly 70 per cent. High self-awareness subjects with few stresses showed a very low relapse rate of about 14 per cent. In the case of low self-awareness subjects, the number of life events was unimportant—relapse rates were about 40 per cent among high- and low-stress sub-groups. In other words, it seems as if the more self-aware people more readily used alcohol to dull their awareness of unpleasant events, while the low self-awareness group had less need to do so. It is interesting to note, however, how well the high self-awareness group did in the absence of stress. This may have been because their increased self-awareness led to better vigilance and, hence, better control over high-risk situations and cues.

If alcohol does serve to dull consciousness and self-awareness, it is not difficult to understand the vicious circle problem drinkers get caught in when they try to reduce their awareness of problems which have arisen from drinking by drinking more. Not only does this sabotage the person's attempts to solve these problems, it also decreases awareness that drinking is a problem in itself. Too often, the dulled mind of the drinker *externalizes* problems; they are blamed on all sorts of things except drinking. This may partly explain why it is so common for problem drinkers to deny their diffculties with alcohol, though many other processes are involved in this denial, as we shall see later. When a problem drinker stops drinking for a few days, or even two or three weeks, a mental dullness persists through the effects of alcohol on the brain. These effects, which continue even though alcohol is no longer present in the bloodstream, are similar in many ways to its immediate effects on the brain. Thus, because of the mental effects of alcohol, a person can be deprived both of an insight into her problems and the capacity to cope efficiently with them. Furthermore, the fact that self-monitoring depends upon self-awareness means that lower-order self-regulatory processes will also be impaired.

Self-image

It is an inevitable consequence of having developed self-awareness that humans acquire a view of themselves in relation to the world, that is, a self-image. That the nature of self-image plays a role in alcohol consumption has been shown by, among others, John Davies and Barrie Stacey of Strathclyde University,[9] who studied the drinking habits and attitudes of a large number of Scottish teenagers. The research concluded that young people are to some extent motivated to drink to avoid the stigma of 'weakness' which is associated in their eyes with abstinence.

How people see themselves will play a large part in determining whether and how much they drink. Similar self-concepts seem to be associated with both smoking and drinking in adolescence,[10] and these self-images include risk-taking and rebelliousness as key components. While risk-taking personalities are more readily attracted to occupations which seem to offer risk and excitement of some sort, and one can therefore expect higher levels of drinking in these occupations, it is also true that jobs can change self-image and hence alter drinking habits. In the case of the merchant navy, the armed forces, and journalism, for example, the role of sailor, soldier, or reporter may influence the person's self-image to such an extent that he takes on the stereotyped attributes of the profession, one of which is hard drinking.

Thus, heavy drinking is not only a consequence of joining a particular occupation or sub-group, it is also a signal to the world about the kind of person one is. Since heavy drinking is inextricably linked with many people's most personal and cherished views of themselves, it is hardly surprising that hard-drinking habits are so difficult to change. However, they do frequently alter owing to natural processes involving changes in self-image associated with the achievement of developmental tasks—in other words, growing up. Possibly the most common 'safety net' for the young heavy drinker who is running into trouble with alcohol in his mid-20s is marriage. The adoption of a new role in society often leads to a fundamental reorganization of the self-image, with the result that heavy drinking is no longer an appropriate symbol. Many thousands of young problem drinkers are 'rescued' from problem drinking every year through similar role changes, though not necessarily all such changes relate to marriage! Job change, educational achievement, religious belief, and joining social movements (CND, political parties, etc.) can all be associated with the adoption of new self-images in which heavy drinking does not play an integral part.

This may account to some extent for the phenomenon of 'spontaneous remission' (see p. 113). When older problem drinkers get over their diffculties, this seems to be related much more to major life changes than to outside help or counselling. While this may be partly attributable to a reorganization of cues and reinforcements for heavy drinking, it may also be due to the change in role, with an associated change in self-image, which usually accompanies major life events like forming new relationships and finding new jobs. Thus, there is an intricate interweaving of conditioning and cognition in the process of spontaneous remission and this is typical of the complexity of all the psychological processes we have been discussing.

Problem drinkers overcome their difficulties in many different ways. Some simply stop or cut down drinking with no fuss or outside help; others do so with the pomp and panoply of a very public religious conversion. There are strong cultural and historical forces providing a 'template' for change which many problem drinkers follow, the most common avenue being through AA. This cultural prescription has, indeed, much in common with a religious conversion experience, and we outlined in Chapter 1 the strong links between the moral–religious movements of the last century and responses to alcohol problems in this. These historical forces are sufficiently strong to influence how the individual behaves when confronting his or her drinking.

One feature of the cultural template is the 'born-again' phenomenon—that iconoclastic shift in personal values, morals, and self-image which involves renouncing one's past, confessing one's sins, and committing oneself to a new way of life. This tradition has had a profound effect on the psyche of the largely Anglo-Saxon culture from which it springs, and fosters a polarization between the 'good' (the saved) and the 'bad' (the damned). As a result of the Temperance Movement's success, the use of alcohol became firmly attached to the 'bad' in the cultural stereotypes of the first half of this century. Throughout northern Europe and North America, there are communities which exhibit this phenomenon in the extreme, the northern Outer Hebrides in Scotland and parts of Northern Ireland being good examples in Britain.

In the context of these cultural trends, it is not difficult to understand the radical changes from drinking to abstinence, and vice versa, which are so common among problem drinkers in Anglo-Saxon cultures. It is equally clear how conflicting pressures from the temperance and drinking lobbies provoke such violent quantum shifts among individuals caught between the competing forces.

Even while problem drinkers are still drinking heavily, it is common for them to go 'on the wagon' for periods. It is as if the oscillation arising from polarization is apparent even in a short time-scale within individuals. The 'Jekyll-and-Hyde' phenomenon is familiar to many problem drinkers and their families, and the possibility that it might not be entirely attributable to alcohol is suggested by the following extract from R. L. Stevenson's *Dr Jekyll and Mr Hyde*:

When I would come back from these excursions, I was often plunged into a kind of wonder at my vicarious depravity. This familiar that I called out of my own soul, and sent forth alone to do his good pleasure, was a being inherently malign and villainous; his every act and thought centred on self; drinking pleasure with bestial avidity from any degree of torture to another; relentless like a man of stone. Henry Jekyll stood at times aghast before the acts of Edward Hyde; but the situation was apart from ordinary laws, and insidiously relaxed the grasp of conscience. It was Hyde, after all, and Hyde alone, that was guilty. Jekyll was no worse; he woke again to his good qualities seemingly unimpaired; he would even make haste, where it was possible, to undo the evil done by Hyde. And thus his conscience slumbered.

This vivid description of a split self bears an uncanny resemblance to the behaviour and testimony of some problem drinkers, and one might speculate that the behaviour of the latter owes more to a potent cultural mythology than to any pharmacological properties of ethyl alcohol.

To return to less speculative issues, favourable attitudes towards the use of alcohol are among the best predictors of whether or not a person will drink heavily. Clearly, where a self-image includes heavy drinking as a key component or symbol, the person will look favourably on drinking. Conversely, where the self-image requires the exclusion of problem drinking because it is symbolically at odds with some underlying attribute, the person will regard drinking unfavourably. Hence the strong association between religious affiliation and light drinking or abstinence which is found, especially in the United States.[11]

One final point about self-image concerns how children whose parents disapprove of drinking react to this disapproval. A higher percentage of such children will end up abstaining from alcohol but, on the other hand, those who do drink will have a higher chance of running into problems.[12] There are a number of possible reasons for this, not least of which is that the children do not learn to drink in a sensible fashion at home. Also, during a time of adolescent development when the youngster is trying to forge an identity distinct from the parents, if alcohol is disapproved of it can easily come to be employed as a weapon or instrument of rebellion. Alcohol used in this way is likely to be drunk quickly, excessively, and in a psychological context which leads to harmful drinking patterns.

Expectations

Different cultures and societies shape how alcohol will be used, and hence how it will be abused, in different ways. They teach their members what to expect of alcohol and how to behave under its influence. The work of McAndrew and Edgerton, who showed how the behavioural and experiential effects of alcohol are shaped more by culturally derived expectations than by pharmacological action, has already been mentioned (p. 240). Some peoples become aggressive with drink, others quiescent; there are tribes who become morose and silent and others who become euphoric. Evidence for this comes also from the psychological laboratories of North America, where an experimental design known as the 'balanced placebo' has been used to disentangle the pharmacological effects of alcohol from those which are learned and determined by expectations (see Chapter 4).

We cannot describe all experiments using the balanced placebo design in detail, but one by Briddell and his co-workers[13] will give the flavour of this type of study. They took a number of male subjects and showed them erotic films, including some involving rape and sadism. The films were shown under each of the four conditions displayed in Fig. 4.2 (p. 124) and sexual responses, in particular penile tumescence, measured. What the experimenters found was that whether or not the young men had actually been given alcohol had no influence on their sexual response to the deviant films. However, when they *thought* they had been given alcohol they became more aroused than when they believed they had not. We have here a good example of how people learn to expect certain things of alcohol. It seems as if alcohol acted as a sort of 'passport' which allowed subjects to respond in ways unacceptable in sober society. Drinking alcohol is associated, in many Western developed cultures at least, with sexuality and aggression, and it is not surprising to find these tendencies released when young men believe that they have taken alcohol, irrespective of whether they have or not.

It is not only in the acute effects of alcohol that expectations have a part to play; they are also important in determining how easily someone who has become dependent will overcome the problem. One particular kind of expectation is known as *self-efficacy*, a term coined by Albert Bandura. This refers to the belief a person has as to whether or not he will be able to carry out successfully a given task. Bandura and his colleagues have demonstrated, for instance, that when a group of snake-phobic students are asked to approach as near as possible to a live snake, the best predictor of how successful they will be is their own advance estimation of their abilities to do so. Bandura distinguishes between *efficacy expectations* and *outcome expectations*. The former consist of beliefs about personal ability, while the latter refer to the extent a given outcome (here, touching the snake) is seen to have desirable consequences for the individual. In conjunction with one of the authors, clinical psychologist Steven Rollnick has applied these terms to an analysis of recovery processes in problem drinkers.[14] Clients' outcome expectations refer to whether or not they believe that stopping or moderating drinking will be beneficial, efficacy expectations to whether they believe they can actually manage to alter their drinking patterns. In other words, change is at least a two-stage process and counselling should attempt to accommodate for this fact. The concept of outcome expectations has many overlaps with the notion of 'expected reinforcement' discussed in the previous chapter (p. 224).

Learned helplessness

Efficacy expectations also have a considerable overlap with the concept of learned helplessness.[15] Martin Seligman, a psychologist at the University of Pennsylvania, has shown in a number of experiments that a predictable psychological state can be induced in animals and humans if they are placed in a situation where what happens to them is independent of what they do. Many of his animal experiments are ethically unacceptable to many people and would not be permissible in the United Kingdom at the present time; however, having been done, the information from them is valuable.

The main way of producing learned helplessness in dogs is to give them inescapable electric shocks. Twenty-four hours later the dogs are placed in a large 'shuttle-box' which is divided into two sections by a low barrier over which the dogs can jump, if necessary. A shock can be given to the dog on either side of the barrier, though the dog can escape or even avoid the shock before it happens by jumping to the other side of the barrier. A signal (light dimming) is given before a shock occurs. What Seligman and his colleagues found was that most dogs which had been given inescapable shocks (no matter where they jumped or what they did they were shocked) did not try to escape when placed in the shuttle-box where they *could* have escaped the shock. This behaviour was apparent in other situations where it was potentially possible to learn to escape from unpleasant stimuli.

Humans were put in equivalent, though less physically punishing situations. One of Seligman's favourite procedures was to induce a state of learned helplessness in subjects by asking them to solve problems which, unknown to them, were insoluble. In this situation, humans and animals tend to develop a particular pattern of responses related to their emotional, motivational, and learning processes. The emotional response is one of listlessness and, in the case of humans, low self-esteem and sometimes depression; the motivational changes mean that people and animals become passive, while human thinking becomes slowed and social functioning handicapped; the learning consequence is that both animals and humans have difficulty in re-learning that their actions can have some effect on what happens to them.

This model can plausibly be applied to a number of situations in the world. The reaction of victims of torture and imprisonment to the uncontrollable environment they are placed in is similar to the listless, passive, dulled picture described above. Similarly, children who have been badly maltreated can show this kind of behaviour, as can the victims of natural disasters such as earthquakes.

How does this apply to problem drinking? There may be some parallel between learned helplessness and the experience of being a problem drinker, at least for those who feel at the mercy of a compulsion to drink. To the extent that the drinker feels out of control of his or her drinking, that person will believe that what he or she does cannot affect what happens. If that belief becomes established, the learned-helplessness syndrome of listlessness, passivity, low self-esteem, and re-learning difficulty is likely to come into play. This picture is one with which those who work with problem drinkers will be very familiar.

What is the solution to the problem of learned helplessness? Seligman and his colleagues 'cured' their dogs by dragging them out of the box in which they were being shocked until they seemed to 'catch on' and re-learn that they could escape. The parallels here with the cue exposure treatments developed by Hodgson and his colleagues (see p. 227) are clear: if you can demonstrate to problem drinkers that they can resist the temptation to drink in situations where they previously found it impossible, you are making the first step towards helping them re-learn that, with respect to drinking at least, they are not the victims of uncontrollable external events. The consequent improvements in activity, self-esteem, and initiative which have been shown to emerge are a major factor in helping problem drinkers—provided always that they see a reason to change.

It is not only with respect to alcohol consumption that problem drinkers may fall into the trap of learned helplessness. Once they have encountered the family, financial, legal, housing, and other social problems so prevalent among this group, it is likely that their difficulties will become so complex, intractable, and interwoven that nothing the person does seems to have any significant effect. Thus, learned helplessness may not be confined to drinking but can become an attitude to life in general and its problems in particular. Later in this chapter we outline some methods which can help drinkers break out of this apathetic trap by teaching them ways of solving pressing life-problems in a systematic fashion.

It is not inevitable, when people or animals are placed in situations where they have no control, that the syndrome of learned helplessness will emerge. An example of this is that roughly one third of Seligman's dogs did not become helpless, in spite of having received the same uncontrollable electric shocks as the others. One reason for this might be that they were to some degree experienced at escaping from unpleasant situations; that is, they had been 'inoculated' against helplessness in the past. Seligman tested this hypothesis by giving some dogs practice at escaping shock before they were given the uncontrollable shocks from which no escape was possible. He found that those who were thus inoculated were much less prone to helplessness in subsequent tests than dogs who had no such previous experience. With regard to problem drinking, one implication of this might be that all those who learn to drink should have some experience of resisting social pressure to drink and of restraining themselves from drinking on occasions when they most feel like it. Thus, they might become inoculated against developing the helpless-type belief that they are victims of drinking circumstance.

Attributions

There is another important element which may determine whether or not a person will learn to be helpless, passive, and listless. Attributions as to the reasons for helplessness made by an individual will be crucial determinants of what conclusions he comes to about his helplessness. Where helplessness is regarded as due to factors which are transient, specific, and located externally to the individual, the experience will probably not have serious effects on self-esteem and motivation. If, on the other hand, helplessness is seen to arise from stable, global, and internally located factors, the effects of being placed in an uncontrollable situation are liable to be more serious.

Consider the case of two men who are learning computer programming on a government retraining scheme. Both are finding difficulty in mastering the necessary skills during the first week; one drops out of the course at the end of the week while the other stays on and finishes it successfully. Why is this? In this example, it was due entirely to the different attributions made by the two men as to the cause of their inability to learn. The man who dropped out attributed it to a lack of academic talent, an attribution which was global (it referred to his entire intellectual capacity rather than his competence at one particular skill), stable (intellectual capacity is largely regarded as fixed), and internal (failure was due to his own deficiencies). The other man, by contrast, attributed his difficulties to bad teaching on the course and a 'rustiness' of his academic faculties. This attribution was transient (new teachers would appear later in the course; his rustiness would go with practice), specific (he did not use this performance as a gauge of his intellectual capacity), and external (his helplessness was attributed to poor teaching and lack of practice, neither of which are personal attributes). In fact, the two men were of similar intelligence, but the first left the course feeling miserable, believing that he could not learn anything difficult and suffused by a listless apathy about making any further attempts to find a decent job. The finisher had his sense of personal competence enhanced by the end of the course; the drop-out had his further depleted.

Again, the implications for problem drinking are many. Where a problem drinker attributes his difficulties to *global, stable* and *internally located* factors, as would be the case when someone attributes problems to his own personal weakness and character deficiencies, it is more likely he will react to his difficulty in controlling his drinking, as well as to general life problems, by becoming passive, listless, and depressed. If he were to attribute his difficulties to *externally located, transient, specific* factors, he would likely show more spirit and fight in facing up to them. An example would be where a person interprets her drinking difficulties in terms of the combined effects of certain cues, reinforcements, and pressures. If this attribution is successfully made, the person will be to some extent protected from learned helplessness. On the other hand, it is essential that she does not abrogate her sense of personal responsibility, for that would be to further sap the feeling of personal control which is at the core of this issue.

Some of the weaknesses of disease models of problem drinking are further revealed by this. The belief that one is the victim of a disease over which one has no control is the epitome of a global, stable belief. It does have the apparent advantage of allowing people to externalize their problems from their personal identities to some extent, but whether this is desirable in the light of the need to assume responsibility for actions is less clear. One obvious disadvantage, though, is that believing problem drinking to be a disease gives the drinker absolutely no help in setting about changing things in a practical way. A more sophisticated set of attributions, whereby the person sees his behaviour as faulty learning primed by cues, reinforcements, and thoughts over which he can potentially gain control, provides a basis for making specific and practical plans.

Abstinence violation effect

Helplessness is such a crucial and pervasive phenomenon that it overlaps into many other realms of the higher cognitive functions. In studying what happens when people relapse after having been abstinent for some time, psychologist Alan Marlatt described the phenomenon of the Abstinence Violation Effect (AVE).[16] However, Marlatt does not confine himself to abstinence from alcohol; rather, he regards the AVE as being a common psychological reaction to the breaking of a personal commitment to abstain from any activity. The constituent elements of the AVE are: (1) an attribution of the 'slip' to one's own personal weakness or failings; (2) cognitive dissonance, that is, the disparity between one's image of oneself as an abstainer and what one now finds oneself doing. Before going on to discuss cognitive dissonance, let us just mention that if failure is attributed to personal weakness, a likely consequence of finding a disparity between self-image and behaviour is that the self-image will change first. Thus, the abstainer who finds herself drinking is likely to throw away the halo of virtue and self-respect and resume the 'I'm a drunk who has no control' self-image. Once this image is re-established, a natural consequence is that the person drinks to live up to it; 'one drink, one drunk' becomes the convenient belief and a vicious circle of learned helplessness and self-fulfilling prophecy sets in. The AVE thus causes a violent oscillation between abstinence and heavy drinking and, with each turn of the cycle, drinkers' beliefs in their inability to control themselves become strengthened and self-esteem and self-efficacy weakened, with corresponding reductions in the frequency and duration of abstinent periods.

One of the methods used by Marlatt to counter effects like these is to prepare the person for relapse, which is something every problem drinker must face sooner or later. So, Marlatt tries to challenge their belief in the 'hanged for a sheep as a lamb' philosophy. Here is an extract from written instructions given to members of a relapse prevention programme:

One swallow doesn't make a summer

A slip is not all that unusual. It does not mean that you have failed or that you have lost control over your behaviour. You will probably feel guilty about what you have done, and will blame yourself for having slipped. This feeling is to be expected; it is part of what we call the Abstinence Violation Effect. There is no reason why you have to give in to this feeling and continue to drink (or smoke). The feeling will pass in time. Look upon the slip as a learning experience. What were the elements of the high-risk situation which led to the slip? What coping response could you have used to get around the situation? Remember the old saying: one swallow doesn't make a summer? Well, one slip doesn't have to make a relapse, either. Just because you slipped once does not mean that you are a failure, that you have no will power, or that you are a hopeless addict. Look upon the slip as a single, independent event, something which can be avoided in the future by the use of an appropriate coping response.[16]

Cognitive dissonance

Marlatt includes many other elements in his relapse prevention programmes which will be described later. Meanwhile, we shall expand on the concept of cognitive dissonance, originated by Leon Festinger, an American social psychologist. When a person holds two incompatible attitudes or beliefs, or behaves in a way incompatible with a given attitude or belief, that person will probably feel some psychological discomfort. A common reaction to this discomfort is to alter one of the beliefs, or one of the behaviours, so that they are more compatible.

Sales and marketing executives use this principle widely; wherever a sales campaign aims to get members of the public to do something active in favour of a product or brand name, it is trying to change customers' attitudes by inducing a state of cognitive dissonance. If people hold neutral or negative attitudes towards, say, Crunchy Corn, the Breakfast Bombshell, but are induced to send off packet tops so that one pence per top will be donated to a worthy charity, it is likely that customers will experience cognitive dissonance. Why? Because their behaviour (going to the trouble to tear off the top, buy a stamp and envelope, and post it off) is incompatible with their attitudes (indifference or mild revulsion to Crunchy Corn). But, having carried out the behaviour, they cannot change it, which leads, in turn, to the attitude to Crunchy Corn changing in a positive direction. For this reason there is hardly a product in existence whose package does not contain exhortations to the consumer to do something in the product's name.

This is not as far from problem drinking as it may seem, for cognitive dissonance has a big part to play in drinking in general and problem drinking in particular. Its relevance to relapse has already been outlined. As far as overcoming drinking problems is concerned, it is possible to understand how cue exposure and other similar treatment approaches might owe part of their effectiveness to the cognitive dissonance set up between a negative attitude to oneself (for example, 'I can't control my drinking') and a particular action (for example, refraining from drinking when suffering from withdrawal tremor).

It may also partly be the case that young people learning to drink develop favourable attitudes to alcohol, owing to cognitive dissonance processes. If they find themselves drinking heavily on a few occasions, say because of a particular milieu such as the armed forces, their attitudes to drinking are likely to fall in line with their behaviour, however much that behaviour is determined by external pressures such as social norms, traditions, and availability. Of course, alcohol advertisers also use cognitive dissonance techniques to mould attitudes to alcohol.

Problem-solving and coping

People view events in very different ways, and how they appraise them will depend on the attributions they make and, hence, on the attitudes and beliefs they develop towards themselves and the particular situation. We have also suggested that life problems, far from being a separate entity from alcohol dependence, are intimately involved in it and that part of the difficulty in changing drinking patterns is attributable to an awareness of life problems on the part of the drinker. For this reason, coping and problem-solving are key factors in determining whether or not problem drinkers overcome their difficulties.

Coping and problem-solving consist of a complex set of psychological processes, ranging from self-awareness, through resisting the temptation to act impulsively, to trying out possible solutions in real life. D'Zurilla and Goldfried[17] carefully analysed the constituent processes of problem-solving and we will use their framework to discuss a number of different psychological mechanisms in the problem-drinking area. There are six constituent processes as described below.

1 Mental set The key to a mental set for problem-solving is some sort of detachment from the immediate problem. There are great individual differences in how people react to crises: there are panickers and there are copers. Copers are less likely to act impulsively and react catastrophically to events. The person who thinks, 'This is going to upset my plans quite a bit' is likely to adjust better to the situation than someone who responds catastrophically by thinking, 'Oh God, my life's in ruins; it's all useless; I can't go on.' The latter may act in a way which worsens the situation, such as by going out and getting drunk, while the former is better placed to make constructive steps towards solving the problem.

Evidence from psychologists Spivack, Platt, and Shure[18] shows that delinquent adolescents are more impulsive when faced with difficult interpersonal situations than non-delinquent adolescents. They tend not to anticipate a number of possible courses of action when faced with choosing how to react to a situation; rather, they impulsively opt for the action which first springs to mind. This is not due to lower intelligence, but to a failure to deploy the higher-level executive processes of the brain. To return to the analogy used in Chapter 6, it is as if the managing director of the brain has fallen asleep or is acting inefficiently. The result of this lack of forethought is that the individual runs into trouble with other people and the law. It may not be a coincidence that delinquency is a major predictor of alcohol problems. Could it be that poor problem-solving ability is the cause of both problem drinking and delinquency? We must not fall into the trap of locating alcohol problems entirely in the individual, but it is interesting that some evidence exists to show that poor problem-solving ability in youth does, sometimes, in itself, predict subsequent problem-drinking.

We have seen how a failure to acquire an appropriate mental set can lead people into trouble, but inappropriate mental sets also cause difficulties. Aaron Beck,[19] a psychiatrist in Philadelphia, and Albert Ellis,[20] a New York psychologist, have both proposed that emotional and behavioural problems can be caused by holding irrational beliefs about oneself and the world. For instance, someone who firmly believes that he must be perfect is almost bound to feel frequently depressed because of the impossibility of achieving perfection in everything in life. Similarly, someone who believes she should be liked by everyone is condemning herself to emotional stresses whenever she encounters someone who does not like her, as must inevitably happen. Ellis writes of the 'tyranny of shoulds and musts' which drive people towards unachievable or unrealistic goals. Alan Marlatt suggests that some problem drinkers' life-styles are so imbalanced towards fulfilling 'shoulds' (for example, 'I should get fit' or 'I should get on better at work') that the only way they know of escaping from this pressure is a heavy drinking session. In this way, one more of the violently oscillating cycles is set up. If this cycle is to be broken, the person has to find ways of escaping from self-imposed demands before they build up to a pressure to drink heavily. Marlatt talks about building a balance between 'wants' and 'shoulds' in creating a new life-style.

Beck has identified a number of logical flaws which pervade the thinking of those with emotional and behavioural disorders. One of these is *over-generalization*, whereby a person takes one bad experience as evidence that something terrible is wrong. So, the over-anxious man assumes that because his boss criticized one report, he must think he is no good at his job. Similarly, the problem drinker who takes a drink one day may assume that this is evidence that she has lost control of her drinking completely. A second common type of error is '*catastrophization*', which means exaggerating the consequences of a difficulty out of all proportion. Thus, the man in the example above concludes that he will get the sack, his life is in ruins, and so on; in a similar way, the problem-drinking woman may conclude that she is helpless, useless, feckless, and worthless. A third logical flaw is *jumping to conclusions*. For instance, the problem drinker who has been abstinent for three months is asked by his wife if he was working late when he comes home an hour late. He infers from this that she mistrusts him and thinks he has been drinking; he flies off the handle and storms out to appease his ruffled pride with soothing whisky. He has arbitrarily taken one possible interpretation of his wife's question, when many other interpretations might have been made.

Beck has also developed a form of counselling known as *cognitive therapy* which aims to change dysfunctional thoughts and beliefs. His therapeutic principles are of considerable potential in dealing with alcohol problems, through helping counsellors identify and alter such thoughts as 'I can't stop', 'I've no control' and 'I can't cope'—thoughts which are all too evident among problem drinkers and may act as self-fulfilling prophesies.

One final issue on the topic of mental set in problem-solving relates to *cognitive control*, that is, the use of various mental strategies to determine how one views a situation. As mentioned, different people can make very different interpretations of the same event, distorting the facts to fit their own expectations and interpretations of reality. This can be put to both good and bad ends; some guards in Nazi concentration camps managed to avoid guilt and compassion for the inmates by denying to themselves the humanity of the prisoners and regarding them as sub-human. On the positive side, political prisoners in many parts of the world have kept their sanity during torture and imprisonment by viewing their gaolers as misguided rather than as malicious ogres.

Earlier in this chapter, we pointed out that immediate reinforcements are very powerful determinants of behaviour. The tendency to go for such reinforcements is a strong, atavistic one, and our adjustment to complex human society requires us frequently to delay immediate reinforcement for some more distant, usually more valuable reward. So, the student has to learn to put off a visit to the pub until after an assignment has been completed, the child must put away the sweets until after the meal and so on. How do we sustain ourselves between the point of resisting the initial urge and achieving the distant reward? How do students sustain themselves through years of study for an uncertain future reward of status and salary, forgoing all sorts of immediate temptations and pleasures on the way? Self-reinforcement and goal-setting, which were discussed earlier, are two means of achieving this. Additionally, however, individuals manipulate the context of their thought, both words and images, to help them transcend immediate rewards and punishments in the pursuit of long-term goals. Various strategies exist for achieving this; Walter Mischel[21] at Stanford University has shown how children gradually develop self-control through employing these strategies. Distracting onself from a reward is one way of reducing the desire for it. One of Mischel's studies showed that children could delay taking a reward ten times longer when they could not see it than when they could. Thinking in a cool and detached way about it also reduces its attraction, in comparison to 'hot' thinking, which is focused on the consummation of the desire. Self-instruction is another commonly adopted means of impulse control in which children, or adults, guide their behaviour through the pitfalls of temptation by instructing themselves verbally, either overtly or covertly, during a sequence of actions.

The implications of these examples for alcohol problems are obvious. The various mechanisms of cognitive control, such as distraction, self-instruction, and altered perception, could all be used by the problem drinker who is learning to overcome the impulse to drink by adopting a more helpful mental set.

2 Problem definition The second stage in problem-solving requires the individual to set out in precise terms the nature of the problem faced. Too often, people lambast themselves with alarming generalities which create a state of anxiety, inhibit constructive action, and obscure the true nature of the problem. For instance, an individual may react to marital and financial difficulties by thinking, 'My life is a mess', 'I can't stand it', etc. More precise problem definition would require the person to specify in detail what the marital and financial difficulties were

('My wife and I are continually bickering over how much I spend on alcohol', for instance).

With regard to alcohol problems, one of the major tasks for clients is to identify in advance the high-risk and problem situations which may lead to relapse. This requires them to spell out in some detail a number of problem situations, so that workable solutions can be sought and rehearsed. A good example of this is some work from Marlatt's research centre in Seattle. A number of high-risk situations were identified in considerable detail for each of a group of severely dependent, in-patient problem drinkers. Three typical situations are as follows:

(1) 'You have worked for some time. Someone, staff or boss or an important customer, criticizes your work and demands that you make certain changes. You think that the criticism is unfair and the changes unreasonable';

(2) 'You are at a party getting to know a girl. Your host asks what the two of you want to drink. Your friend asks for a cocktail and you ask for a soft drink. She turns to you and says "What's the matter, don't you drink?"';

(3) You are sitting alone at home on a Saturday night and you hear the sounds of a party coming from a neighbouring apartment.

These are all examples of clients specifying in some detail the sorts of situations which were liable to be problematic in the future; the next task is to identify potential solutions.

3 Generation of potential solutions The advantage of precise problem definition is that a larger difficulty may be broken down into constituent parts, the potential solutions to which will not appear so difficult. The pessimism, low self-esteem, or learned helplessness of problem drinkers may mean that they fail to see any workable solutions, and this makes it doubly important to break down problems into smaller parts. The problem situations given above are examples of this and, while there are no easy answers to any of them, at least in a supportive relationship with a group or a counsellor, a person might begin to think of a few potential solutions to each of the problems in turn. 'Brain-storming' can be useful here, both as a means of loosening up thinking and as a way of overcoming pessimism. The rules for brain-storming are simple:

(1) any idea, no matter how ridiculous, is acceptable;

(2) do not censor ideas—just spit them out;

(3) no idea is to be mocked or scorned—the ridiculous is encouraged;

(4) try to let others' ideas spark off ideas of your own.

Members of the group, even if it is only two people, then fire out ideas in turn until a long list of potential solutions, more or less practicable, is collected. (This method has been used by us to find alternative pursuits to drinking for young offenders with alcohol-related problems.[23])

4 Decision-making After the solution-generation stage, obviously unworkable ideas are discarded and a short-list of possible solutions drawn up. There is no easy way to come to a decision, except by discussion with a number of people as to the relative merits of various courses of action. In some cases, it may be useful to carry out a simple arithmetical procedure to calculate the 'subjective expected utilities' of each solution; if someone is agonizing over a decision, this method can help systematize thinking.[24]

5 *Verification* In trying out a chosen solution, the subject can simply talk through the likely course of events with friends or counsellors, carefully go through the events in imagination, role-play the solution where appropriate, or actually go out and enact it with the support of someone else. In many cases, all four methods will be used. The Seattle researchers had their group of clients role-play solutions to problems after they had discussed them. The situation would be read to them on an audio-tape and they had to come out with a course of action, including how they would respond to high-risk interpersonal situations. One interesting finding was that the latency of response (how long it took clients to arrive at an answer to a particular problem) was a significant predictor of how well they were doing six months later. Was this because the failures were mentally slow, and less able to 'think on their feet' and cope with pressures and problems as they arose?

The importance of role-playing—of doing as compared with talking—cannot be overemphasized with problem drinkers, at least half of whom will be suffering from mental slowness and thinking and memory difficulties for weeks, months, and even years after they have stopped or cut down. This is particularly true for drinkers over the age of about 40, for whom reversibility of cognitive deficits is much less likely.[25] Social-skills training, drink-refusal training, assertion training, and problem-solving skills training[26] are all examples of 'doing' therapies which seem more effective than the more traditional 'talking' therapies. Goal-setting, practical homework, and tasks such as those set for clients by Blakey and Baker (p. 226) are all potentially useful methods very much in the spirit of active therapy. Concrete actions are learned better than abstract concepts among normal people; for many problem drinkers, their cognitive deficits are such that, in the short- and medium-term at least, it may even be a waste of time dealing in abstract concepts. It is no coincidence that the programme of AA is so concrete, repetitious, and highly structured.

6 *Feedback* Without feedback as to performance in every sphere of our lives, we would give up doing most of what we do in the way we do it. We need to know from colleagues, families, and friends how we are doing as workers, spouses, and companions; marital breakdown all too frequently ocurs because one partner ceases to bother expressing affection and love for the other, or showing gratitude for what he or she does. Feedback and reinforcement overlap considerably and in the same way that many problem drinkers are 'reinforcer less', so many are 'feed-back-less'. Wherever a solution to a problem is tried out, clear, frequent, and predictable feedback is necessary. For the problem drinker who feels helpless and has low self-esteem, that feedback must be immediate, strong, and frequent to help him break out of the cage of learned helplessness. In practical terms, a counsellor can provide this feedback/reinforcement but, to increase the chances of success, significant people in the person's life should be included, especially the spouse.

Where a particular solution turns out not to be working, feedback may have to be negative. So the decision to seek marriage guidance may lead to no improvement in the home situation. So sensitized are many problem drinkers to repeated failure, they may react catastrophically to yet another, and all the resources of counsellors and friends will be needed to discourage the person from engaging in dysfunctional thinking, which leads to further drinking and 'giving up'.

9 Some practical implications

No one theory is absolutely true for all time. Newtonian physics appeared to explain most physical phenomena satisfactorily until Einstein developed the theory of relativity. For most situations on earth, however, Newton's theory of physics allows us to make accurate predictions about space, time, and motion, and is still useful within clear limits. If it is their usefulness which distinguishes good theories from bad in the physical sciences, then in the behavioural sciences, like psychology, this is even more true. The touchstone for evaluating the new paradigm of problem drinking must be how useful it is in understanding and predicting the behaviour of problem drinkers, as well as in suggesting effective means for solving their problems.

For several reasons, it is not possible for psychology to lay out a set of well-tested axioms and corollaries about human behaviour which come anywhere near the rigour and precision of the physical sciences. Theories abound regarding one small aspect of human behaviour—problem drinking—yet many are merely fashions which provoke a flurry of anecdotal case reports in the professional journals before disappearing into oblivion. These fashions do a great disservice to the problem drinker because they tend to induce a heady optimism and zealous commitment to the theory and its associated therapy which lasts only as long as the enthusiasm itself. Once the optimism of staff working in this exceptionally difficult area has been exhausted, the old pessimistic 'nothing works' belief is disinterred, and cynicism and 'burnt-out' disillusionment are reinforced. In this climate of theoretical and therapeutic nihilism, deeply entrenched moralistic views of problem drinking can all too easily come to the surface. When professionals who work with problem drinkers begin to moralize, the chances of a more enlightened public attitude are much reduced.

So, is social-learning theory a fashion which will pass in a year or two, leaving a rump of moral opprobrium for the problem drinker as its legacy? The answer, for a number of reasons, is most emphatically 'no'. First, several treatment methods stemming from social-learning theory have been subjected to controlled, scientific evaluation with promising results; second, at least some of the tenets of social-learning theory have been scientifically tested; and, third, no other theory in the field of alcohol problems has generated as much empirical research or made as many predictions about effective treatment and prevention. As the pioneering social psychologist Kurt Lewin once remarked, 'There is nothing so practical as a good theory.' We will consider some practical implications of social-learning theory shortly.

One point should be made before moving on to these implications. It is a fundamental principle of social-learning theory that the abnormal is quantitatively, but not qualitatively, different from normal, and we have taken pains to relate aspects of normal drinkers' use of alcohol to the experience of problem drinkers. If there is a continuum between normal and problem drinking, it follows that intervention should also be on a continuum between, at one end of the spectrum, treatment of the individual case and, at the other, large-scale prevention. The notion of a continuum of treatment[1] has consequences for a number of types of intervention with people whose drinking is not yet serious enough to warrant seeking help at a conventional clinic, yet nevertheless causes problems. These secondary-prevention strategies will be discussed in a separate section. First, we consider the implications of social-learning theory for the types of client who have traditionally appeared at alcoholism clinics.

Treatment implications for clinic attenders

Were clinics to operate on social-learning-theory principles, there is reason to believe that clinic-attendance patterns might well change. One reason for believing this is that the possibility of controlled drinking as a viable goal for some problem drinkers follows directly from a social-learning model. If it were widely known that giving up alcohol completely was not necessarily a prerequisite for overcoming all drinking problems, it is likely that more people would seek help and, in particular, that more people with less severe problems would seek help. The improvement in success rates which would probably result would boost staff morale, improve the public image of treatment agencies, and perhaps attract yet more referrals. If this were associated with an abandonment of the 'psychiatric' associations of such clinics and with the adoption of a less intimidating, 'educational' ethos, it is possible to envisage a snowball effect of improved public image, increased referrals, and reduced stigma.

There are more wide-reaching implications of the rejection of a disease model of alcohol problems, however, since the way in which the problem is conceived at the individual level influences political and social reactions to it. Take two contrasting examples, Finland and the United States. In Finland, problem drinking is regarded as a social phenomenon and social workers are the main group dealing with it; the medical profession restricts itself to treating the physical consequences of drinking and psychiatrists have little to do with problem drinkers. In America, problem drinking is seen predominantly as a disease and psychiatrists are the major force in treatment centres. In contrast to Finland, treatment is largely conducted in expensive in-patient settings where the veneer of 'medical treatment' is highly polished. Expensive private clinics vie with each other to attract patients through glossy television commercials.

It is perhaps no coincidence that there is much greater public acceptance of legal and fiscal preventative measures for alcohol problems in Finland[2] than in America, and that the former country has lower rates of most types of alcohol problem. For, if problem drinking is viewed predominantly as a disease, located in the individual and treated by the medical profession, why should the public accept the need for fiscal and legal measures to curb it? If, on the other hand, drinking problems are viewed as social phenomena, there is every reason for the public to accept social and political measures. Of course, other factors complicate this argument, not least the state control of alcohol production and distribution in Finland. We are simply suggesting that a prevailing medical conception of alcohol problems may detract from the development of effective prevention policies. Certainly, it seems to be the case that the dominance of the disease model for the last twenty years in Great Britain has held back imaginative thinking.

Treatment cannot be divided from prevention policies, and treatment ideologies are intimately connected with social policy. We shall now investigate treatment and prevention policies in the light of a social-learning model of alcohol problems. The starting point is to ask whether effective 'treatment' of individual, self-referred problem drinkers is possible.

Does treatment work?

There were two studies in the last decade which have been enormously influential in their effect on thinking about the treatment of alcohol problems. In the first, a review of over 384 treatment-outcome studies, American psychologist Chad Emrick concluded that there was no evidence that treatments of any length produced better results than no treatment or very minimal interventions.[3] The second study, carried out in

London by Jim Orford and Griffith Edwards,[4] compared simple assessment and advice with comprehensive treatment in a group of 100 married problem drinkers. At a follow-up, two years later, there appeared to be no difference between the two groups in how they were faring.

These findings found echoes in the work of researchers who had looked at what made problem drinkers change their drinking habits, whether towards controlled drinking or abstinence. As mentioned earlier (pp. 113, 234), major life changes in work, relationships, housing, and other areas were associated with improvements in drinking. In addition, many people report sudden, inexplicable changes in the way they see themselves and their behaviour, leading to radical alterations in drinking. While few recovered clients in the Orford and Edwards' study credited treatment with having had much of an influence on their improvement, the advice and assessment they had received was accorded a high place in the list of influential events. Emrick's work does not contradict this, for he found no difference between more intensive treatments, on the one hand, and minimal treatments, on the other. It may be, therefore, that in the course of reviewing her life history and current situation in the presence of an empathic yet neutral outsider, the problem drinker gains a perspective on her conduct which is new and salutary. When, in addition, clear and specific advice is given about drinking, it may be that, for some people at least, a brief intervention has an influence disproportionate to its duration. This was the view of Orford and Edwards.

So, do problem drinkers (including 'alcoholics') need no more than good assessment, clear advice, and regular follow-ups? The answer is negative, at least for a proportion of problem drinkers. It is true that for those who have stability of family, work, and friendships, this may be all that is needed; indeed, more may actually be harmful, because it may sap a sense of personal responsibility for their actions and may shield them from the consequences of their conduct. But for other problem drinkers, more help may be necessary.

How can this assertion be reconciled with the findings of Orford and Edwards and of Emrick? First, treatment based on a social-learning model simply has not been tested out to anything like the same extent as conventional treatments. Second, in a few studies where such methods have been tested, positive and promising findings have emerged, as is shown by the work of Chaney, Marlatt, Hunt, and Azrin and several others (see Chapters 7 and 8). We shall now summarize the most recent evidence about what might work and what might not, before making some suggestions about the development of treatment policy in Britain.

A recipe for effective counselling

There follows a list of nine ingredients in a recipe for effective counselling, some based upon empirical evidence and others more theoretically grounded. The nine ingredients are:

(1) understanding and treating drinking in context;

(2) problem-solving;

(3) 'more action, less words';

(4) family, friend, and community involvement;

(5) 'make it worth their while';

(6) mastering the cues;

(7) self-management and target-setting;

(8) counsellors and models;

(9) client choice.

1 Understanding and treating drinking in context At the risk of
labouring the point, drinking alcohol is not a thing apart, a
chemically primed and mysterious urge that sweeps aside the
laws of behaviour. This is as untrue for the 'alcoholic' as for the
normal drinker. Most problem drinking is intelligible in terms
of the social-learning principles outlined in Chapters 6, 7, and
8. It follows that it is no longer enough to say, 'This person is
an alcoholic and must stop drinking'.

Drinking must be understood in its relationship with the
drinker's environment (cues and consequences), emotions
(anxiety, depression, anger, grief etc.), attitudes (beliefs, self-
image, expectations, attributions etc.), and social relationships
(assertion, social skills, marital conflict, sexuality, role models
etc.). It may be because the heterogeneity of problem drinkers
has not been properly recognized, and appropriately differen-
tiated treatment strategies developed, that pessimistic conclusions about treatment effectiveness have been arrived at. No
one 'treatment programme' can be expected to cater for the
needs of all problem drinkers, yet such monolithic enterprises
have been the rule rather than the exception. Programmes such
as problem-solving skills training (see p. 267) and community
reinforcement (see p. 231) are far from uniform; in both, the
content of treatment is based on an analysis of the context of
drinking for each individual. Training in problem-solving skills
requires the person to identify personal problems and difficulties and then rehearse possible solutions. Similarly, community-
reinforcement procedures tailor their interventions around the
personal circumstances of each individual. It is a necessary
condition for any effective treatment that it understand and
analyse drinking in the total social, environmental, and psychological context, and from there design an appropriate
intervention.

If problem drinking is to be understood in context, it must

also be treated in context. From a social-learning viewpoint, the most absurd way to treat problem drinkers is to take them off to a residential institution miles away—physically or psychologically—from their home community. A drinking problem does not reside in the individual but arises from the interaction between the individual and his environment. If drinkers are to resist cues for drinking they must be in contact with the environment which conveys these cues; if they are to lose conditioned tolerance, they must face up to the stimuli of the environment in which tolerance has developed; if there is to be a change in the incentive or 'pay-off' balance in favour of abstinence or controlled drinking, that change has to be negotiated and arranged in the drinker's family and social environment. It is not much use for the person to undergo a change of self-image in the rarefied atmosphere of a residential setting, for self-image is determined largely by the everyday world around us. Pious hopes and earnest resolve frequently last no longer than the bus ride home.

It is nevertheless true that there is not enough residential accommodation for problem drinkers. Many are homeless or living in unsatisfactory conditions, and decent accommodation is a prerequisite for change. But accommodation and the provision of the physical and emotional security of a home is not to be confused with the treatment of alcohol problems in those who already have reasonable homes. Residential care is a pressing necessity for problem drinkers with accommodation problems or other social difficulties, and may also be necessary for the treatment of some medical conditions associated with heavy drinking. However, to the extent that residential care extracts the drinker from the environment to which he must return, it will be counter-productive unless there are good practical reasons for it.

2 *Problem-solving* We have seen that alcohol problems have more to do with the tangles and difficulties in the drinker's life

than with a dependence on the drug ethyl alcohol. It follows that a central feature of any counselling should be the systematic tackling of these problems. While everyone who works with problem drinkers would rightly claim that such problem-solving was a fundamental part of their job, relatively few will go about it in the systematic way described in the previous chapter.

The reasons why problem-solving is not best left entirely to common sense lie in some basic psychological principles. One relates to success and failure experiences and the need for frequent feedback about achievement. For many problem drinkers, their difficulties appear so overwhelming that they are disabled by a suffusing mental 'helplessness', which may take skilled counselling to overcome. One of the chief ways this can be done is by breaking down the problem into its constituent specifics and, from there, actively rehearsing and practising potential solutions. The principles which determine whether or not an individual will learn to tackle problems appropriately can be applied with greater or lesser effect; some counsellors use them instinctively and systematically, and some do not use them at all. This may be one factor underlying the obvious fact that some counsellors are more effective than others.

3 'More action, less words' To an extraterrestrial, the practice of most traditional in-patient alcoholism clinics would appear bizarre. Men and women are whisked away from where they live, persuaded to talk about themselves for a few weeks, and then whisked back home to the same environment. From a Martian perspective, it is little wonder that treatment is commonly believed not to work! What is supposed to have changed? Attitudes, beliefs, emotions? These may indeed change in the rarefied and protected environs of the clinic, but the chances of their remaining changed once back in the company of old family friends and neighbours are slim. American soldiers subjected to years-long political brainwashing in

captivity by the North Koreans quickly reverted to their original beliefs and attitudes on return to their homeland, even though while in captivity they were convincingly mouthing the Marxist slogans of their captors. Why was this? Because beliefs do not exist in a vacuum; they are attuned to prevailing social norms. If the North Koreans could not create a lasting change in beliefs over a period of years, how can alcoholism units expect to do so in a matter of weeks? Changes in beliefs, attitudes, and behaviour need the support of the social and physical environment if they are to be anything more than ephemeral.

This brings us back to problem-solving. Even the most alert and intelligent human being finds it hard to solve practical problems in the abstract—they must be tested out in practice! With a group of alcohol-befuddled individuals, some of whose intelligence, concentration, and memory will be impaired for many weeks or, in some cases, for many years, to expect their abstract deliberations about physically, emotionally, and temporarily distant events to have much significant carry-over to future behaviour is bordering on the absurd. For there to be any such carry-over, all our knowledge about learning would have to be put into practice. It would require frequent repetition of what is to be learned, abundant practice, revision, and, perhaps most importantly, involvement of those family and friends who will be in contact with the 'trainee' to ensure that what has been learned is maintained once the 'course' is over.

Seen in this light, the relative effectiveness of treatments emphasizing action is less surprising. Social-skills training, problem-solving skills training, community-reinforcement programmes, cue-exposure regimes, and graded exposure procedures (see Chapters 7 and 8) all require the client to do more than talk and this is a necessary condition for effective treatment for all but the most socially stable clinic attenders.

4 Family, friend, and community involvement Social stability, or the extent to which someone has solid and enduring roots in family, friends, and work, is one of the best predictors of who will solve an alcohol problem and who will not. While it does not follow directly from this that there is a causal connection between social stability and outcome, there are strong grounds for supposing that such a link may exist. If social forces naturally help people to overcome drinking problems, a counsellor might profitably harness the same forces.

Marital counselling does seem to be an effective means of helping some problem drinkers.[5] Whether involvement of the wider family, friendship, and community networks similarly improves treatment results remains to be seen, though from a social-learning point of view there are good reasons for believing that it would. As mentioned in the discussion of 'more action, less words', our behaviour is determined as much by what happens around us as by what happens within, though external and internal events influence each other strongly. What events could be more important and influential than the behaviour of people close to us? The complex family dynamics which emerge in alcohol-complicated marriages cannot be discussed here,[6] but the point being made is that the family should, wherever possible, be involved at every stage of counselling.

It is not only the family which may become involved. Suppose a counsellor advises a male client that he should avoid drinking alcohol after work. How much more effective this advice would be if the client had agreed to bring along one of his workmates whom he trusted and liked, and if this friend became a party to the advice and was ready to help the client out during crises and difficulties! Many problem drinkers will not have such helpful friends, but it is possible to find some appropriate friend, family member, or workmate for a good proportion. (One of us has succeeded in persuading a young, problem-drinking delinquent to bring along members of his gang, who subsequently kept their friend strictly on the path of controlled drinking because his drunkenness was liable to get *them* into trouble!)

The way other people become involved during counselling will depend on the therapeutic orientation of the counsellor but an obviously important constituent is that there should be warm and positive human relationships between those involved. (For example, a husband may have difficulty in dissuading his wife from drinking on a particular evening, even though this had previously been negotiated by the counsellor, if their relationship is fraught.) This is as true for hostels for problem drinkers as for families and friends, and it is the counsellor's job to foster such good relationships.

This is not by itself enough, however, for a problem drinker is still responsive to cues and consequences, faulty expectations, and disruptive attributions. Family and friends can become a party to a programme of cue exposure, graded exposure, stimulus control, instrumental learning, and all the other social-learning based strategies, provided this is carried out within supportive and warm human relationships.

5 *'Make it worth their while'* One of the major tasks for those close to a problem drinker is to develop a consistent policy towards that person's drinking. Relatives and friends often unwillingly reinforce heavy drinking in their desperation to do something. The wife is censorious and avoiding during one week, guilty and forgiving the next; she tries everything, but nothing works; meanwhile the drinker's friends are consoling him with pints of beer in the pub and 'curers' the next morning when he arrives shakily at their door for commiseration after a row with the wife.

Whatever the origins of the drinking problem, recovery requires that drinkers are brought face-to-face with the consequences of their behaviour. The worst thing a relative, friend, or workmate can do is to collude with or cover-up for the drinker, and bad treatment can also act in a similar counter-productive way. Many drinkers have stopped or cut down because their wives or husbands have left them, because they have received a final warning at work, or because they have been arrested by the police. Obviously, many have not, though this does not argue against the usefulness of building consistency into how people respond to the problem drinker's behaviour.

In Hunt and Azrin's 'community reinforcement' approach (p. 231), the whole environment of a group of problem drinkers was reconstructed so that abstinence was rewarded and drinking led to exclusion from rewards. Although this scheme is too expensive for routine application, it principles could profitably be applied within routine services. Exasperated friends, family, and employers might be prepared to re-extend human companionship to the problem drinker on the basis that he must only see them in a sober state. The guilt and resentment which arise under normal circumstances from turning away a drunken friend or relative would be much ameliorated if everyone concerned, including the drinker, had sat down and negotiated a contract with the help of the counsellor. The contract could then stand as a shield between them and emotional blackmail or manipulative behaviour by the problem drinker. After an initially difficult period, it may well transpire that, if the contract is consistently adhered to, the drinker learns that he is not being rejected as a *person* but that it is his heavy drinking *behaviour* which is being rejected. So long as friends or family do not reject the person for drunkenness outside their presence and assuming they avoid a moralizing and censorious stance, it is possible that their friend's or relative's excessive drinking would at least cease for periods when he wanted their company. That would be a worthwhile start.

The important point is that the person must see some reason to change. Making it worth their while does not entail punishing them for drinking; it means helping them make life more pleasant when abstaining or drinking lightly than when drinking heavily. It also means helping them ensure that life does not conspire to be sweeter under the influence of large amounts of alcohol, though for many of the friendless and deprived this position is exceptionally difficult to achieve.

6 Mastering the cues As shown in Chapter 7, drinking is, to some extent at least, a habit. Habits are mainly regulated by cues, or stimuli, on the one hand, and consequences, or reinforcements, on the other. We have just seen how consequences may be altered to change drinking habits, and Chapter 7 gave a number of examples of how cues may be altered, avoided, or neutralized in the attempt to alter the drinking habits elicited by these cues. Stimulus control and cue exposure were two of the methods discussed.

While many problem drinkers overcome their problems without ever identifying the stimuli which elicit their desire to drink, there is no doubt that many do adopt such methods spontaneously or that programmes using these methods can be successful.[7] It is well within the capabilities of most counselling services, given some training and supervision, to implement stimulus-control procedures and some types of cue-exposure programme. The procedure devised by Blakey and Baker (see p. 226) is a good example of the sort of graded exposure programme which is within the scope of many counselling agencies. It allows counsellor and client to set specific targets, and for both to experience an increase in self-esteem as each sub-goal is achieved.

In many cases it may be necessary to avoid certain cues, rather than neutralize them through cue exposure. The first step here is self-monitoring (p. 241), a means by which drinkers discover which cues are associated with heavy drinking. Although self-monitoring has typically been used where people are being encouraged to cut down, it is also useful when abstinence is the goal, since clients can monitor in detail the circumstances when they experience a strong desire to drink, as well as those when they succumb to temptation. On the basis of such records, guidelines can be negotiated so that 'high-risk' cues are either avoided or, at least, the problem drinker is well-prepared in advance for them. For instance, in a case where being alone at the weekend is a cue for drinking, counsellor and client may write out an 'emergency telephone card', with enough telephone numbers of potentially sympathetic people (for example, hospital clinic, AA, counselling service, brother or sister in London, or volunteer counsellor) so that the client is able to resist 'loneliness cues' with the support of a friendly voice on the telephone. That is one simple example; ingenuity need know no bounds when planning cue-coping strategies. Such self-regulation procedures are also readily self-administered and form the corner-stone of many self-help manuals.

With regard to the more complex cue-exposure programmes in which cues are internal ones associated with waning blood alcohol levels, the types of procedure carried out by Hodgson, Rankin, and Stockwell (pp. 227–8) could not easily be administered in the typical counselling service in the absence of specialized supervision. These treatments are still in the early stages of development, but such are the taboos about giving alcohol to problem drinkers that hardly any research is being carried out in the UK in this area. If withdrawal symptoms can be de-conditioned in the manner described by Hodgson and his co-workers, this has profound implications for the treatment of drinking problems. It is most unfortunate that so little of this research is being conducted.

7 *Self-management and target-setting* Whenever human beings embark on some enterprise which entails a deviation from habit and routine, and particularly if the project is liable to be arduous and sustained, certain mental processes can be identified which are prerequisites for success. The first is the deployment of attention to onself and the environment, in other

words, self-awareness or self-monitoring. The second is the setting of targets, or sub-goals, leading to the final target. Such sub-goals are necessary because human beings crave feedback as to their performance, and this serves both an informational and motivational (or reinforcement) function. Third, humans need to gain satisfaction from what they are doing if they are to sustain effort over long periods and must find ways to reward themselves, mentally or physically, for achieving targets. They must also use a variety of mental strategies (distraction, 'cool thinking' and so on; see pp. 270–1) to steer themselves away from immediate temptations.

To suggest that self-management procedures should be central to alcohol counselling is not to advocate an ephemeral therapeutic fashion. While overcoming an alcohol problem can mean different things to different people—changing an identity, giving up a value system, being confronted with hitherto avoided realities, and so on—what it means for every problem drinker is overcoming a more or less entrenched habit. To make possible a change in habits, self-monitoring, target-setting, self-reinforcement, and all the other self-regulation strategies are required. Fortunately, therapeutic procedures exist which are designed to strengthen these self-regulatory processes.

Most people develop their own self-management strategies, but some do not and many do not have access to the full range of available procedures. Where a habit is powerful, the drinker should possess information about all possible options so as to build on personal strategies he has already developed. These strategies are outlined in self-help manuals and their presentation in a clear, non-jargon form may be a useful adjunct to personal counselling for the drinker with serious problems and a strongly developed habit.

8 Counsellors and models Some counsellors are better than others at helping problem drinkers.[8] One of the features distinguishing more from less effective therapists is empathy, the ability to communicate an understanding of how the client thinks and feels about his or her problems. Many problem drinkers find that talking to a sensitive listener is enough to help them work out solutions to their difficulties and build up a determination to do something about them. If counsellors are to be equipped to deal effectively with the broadest range of problem drinker, however, they must be able, not only to listen emphathically, but to act as advisers, role models, coaches, and environmental engineers. They should be able to run a social-skills training session, carry out a functional analysis of drinking, negotiate targets and drinking guidelines, plan a cue-exposure programme, draw up a behavioural contract between drinkers and their spouses and/or friends, and so forth. Otherwise, they will be limited in the number of people they can effectively help.

No study has ever shown professionals, whether psychiatrists, psychologists, social workers, or nurses, to be any more successful at helping problem drinkers than non-professionals given a brief training in alcohol counselling. Perhaps this is not surprising, given that the main therapeutic means of effecting change in the past has been empathy, for professionals have no monopoly on this. If there is something in the therapeutic methods derived from social-learning theory, then competence at carrying out problem-solving procedures, assertion training, graded exposure and other related methods should enhance counsellor effectiveness. It is competence at these procedures which should influence counsellor effectiveness, rather than professional training, most of which does not include any specific training in these skills.

It may be, however, that research will prove this hypothesis wrong. Perhaps life-events, housing, the making and breaking of relationships, work, and all the other factors which influence drinking will continue to overhadow the modest effects of individual counselling. Perhaps within these individual counselling relationships, empathy will continue to dominate technical expertise. But this must be put to the test before writing off all specific methods as adding nothing to the fundamentals of counselling.

9 *Client choice* Given the relative ignorance about the nature, causes, and cures of alcohol problems which has prevailed over the last fifty years, it is somewhat alarming to reflect that problem drinkers have had little or no say in choosing how they should be helped. The dogma of total abstinence for everyone is only one example of this. It has been the rule rather than the exception for problem drinkers to be processed through a standard treatment 'regime', as if they were suffering from a sort of mental appendicitis for which there is some well-established medical treatment.

The bewildering variety of ways in which alcohol is used by people, and the range of ways it can become a problem, is evidence enough that different problem drinkers need different kinds of help. When problem drinkers are allowed a choice, the success rates of treatment tend to improve.[9]

The decision to aim for abstinence or controlled drinking is one example of client choice and, while we cannot go into the complexities of this issue here, it can be said that between 15 and 37 per cent of in-patient problem drinkers would opt for controlled drinking. Among less severe problems appearing at out-patient clinics, this proportion may be somewhat larger. Controlled drinking is definitely possible for many problem drinkers, though for the most severely dependent, older individuals (those corresponding to the stereotyped 'alcoholic' who is drinking very heavily almost every day), controlled drinking is far more difficult to achieve and most would be better advised to abstain, in the short and medium term at least. (We advise anyone who has an interest in the subject of controlled drinking to read our earlier book,[1] in particular Chapter 8, before embarking on any controlled-drinking initiatives.)

When offered a choice of treatment and of goals, it is likely that a greater number of people will be attracted to seek help. In particular, attendance rates might increase among those with less severe problems whose prognoses are so much better, simply because their habits are so much less entrenched. Less severe problem drinkers can, and do, reduce their drinking successfully and insisting on abstinence can actually cause them to drink more in the short term. Martha Sanchez-Craig and her colleagues[10] carried out a study in Toronto with a group of seventy 'early-stage' problem drinkers who were drinking on average about 70 units (1 unit = 1 single measure of spirits) of alcohol per week and who reported significant problems with drinking. Clients were randomly divided into an abstinence and a controlled-drinking group, but the latter were asked to abstain temporarily during the first four sessions of treatment. During the first three weeks, the abstinence group drank much more than the controlled-drinking group and significantly more of the controlled-drinking group actually abstained! Follow-up a year later showed that there was no significant difference between the groups but the abstinence group had had to seek help during the previous year more often than the controlled drinkers.

This and other evidence about the deleterious effects of the abstinence goal for some people constitutes further grounds for making controlled drinking widely available as a treatment goal. However, the fact that controlled drinking is most applicable to those with less severe problems should not obscure the fact that, if a client with more severe problems wishes to attempt controlled drinking and provided there are no medical reasons why not, then the counsellor should attempt to help him or her in this way—assuming always that the client has been informed that this may be much more difficult to achieve than abstinence.

Within the United Kingdom, there is now a consensus that controlled drinking is a desirable goal for those with less severe problems, and controlled-drinking treatment is fairly widely practised in National Health Service clinics.[11] While there is still controversy about controlled drinking for the more typically 'alcoholic' clients, the situation here is much more enlightened than in the USA, where controlled drinking is still taboo in most quarters[12] and research in this area is virtually non-existent.

A framework for treatment

To conclude this section, in which we have outlined the ways in which individual treatment of self-referring problem drinkers should develop, we shall make a final point. This concerns the fact that individual treatment reaches only a small proportion of those with alcohol problems. Potential clients may never be identified by general practitioners, social workers, or other 'front-line' professionals; those identified may not appear for further appointments, or may drop out at any stage of assessment and counselling.

In short, only a minority of problem drinkers ever get 'treated'. One could speculate about the reasons for this; physical distance from the clinic and waiting time have both been identified as factors which decrease attendance rates.[13] But the stigma of presenting oneself to strangers as a person who has 'lost control' of one's drinking, especially in a climate where moralistic and punitive attitudes among professional staff may be encountered,[14] must surely have a large part to play. Where treatment is delivered in psychiatric hospitals in association with mentally ill patients, the stigma must be increased. It is simply unconvincing to the majority of the public and professionals that problem drinking (or 'alcoholism') really is a disease, and nowhere is this clearer than in the latest

Mental Health Act for England and Wales, where alcoholism is no longer considered to be an illness, mental or otherwise.[15]

The consequences are that we must look more closely at the means by which treatment services are delivered, as well as at the ethos and ideologies of the staff who deliver them. The 'treatments' arising from social-learning theory are, in fact, closer to 'education' than treatment and arouse different expectations and behaviours on the part of the staff concerned compared with treatments based on medical conceptions. The medical model has little to offer those trying to help problem drinkers other than to 'persuade them to stop drinking' and 'hold on until we make the pharmacological breakthrough in the treatment of this disease'. In such a theoretical vacuum, is it surprising that the staff of traditional treatment agencies sometimes resort to naïve moralizing in their dealings with clients? Quite simply, there is no alternative!

In other words, a necessary precondition for any treatment policy is that the staff carrying out the treatment should be given a language, a framework of analysis, through which to understand the behaviour of problem drinkers. Until now they have had nothing except a mish-mash of ill-articulated biology, psychodynamic theory, AA folk-science, and nineteenth-century moralizing. It has not served them well.

Alternative ways of delivering help and advice

In the spring of 1983, advertisements were placed in a number of newspapers in Scotland, local and national, inviting readers who thought they were drinking too much to write for free advice. The advertisement is shown in Fig 9.1.

A total of 785 people replied and were randomly allocated to two groups, one of which received a general advice-and-information booklet about alcohol problems, while the other

DRINKING TOO MUCH?

If you have the feeling that you are drinking more alcohol than you should, you can get free help and advice through the post now.

This is not for alcoholics, but for people who genuinely feel they should cut down on their drinking and are having some difficulty doing so.

The advice which will be sent to you is absolutely free. This is not a trick to try and sell you something. It's an official health education project to help people reduce their drinking.

If you think you would benefit from drinking less, just write your name and address in the space provided below, cut it out and send it to: SCOTTISH HEALTH EDUCATION GROUP, FREEPOST, DUNDEE DD1 9XW. No postage is required if the word FREEPOST is clearly written on the envelope.

Please note that your name and address will be kept STRICTLY PRIVATE AND CONFIDENTIAL. No unauthorised person will ever know that you have responded to this advertisement.

I would like advice on how to cut down my drinking.
SEND TO: SCOTTISH HEALTH EDUCATION GROUP, FREEPOST, DUNDEE DD1 9XW. (Block letters please)

NAME: (Mr/Mrs/Miss) ______________________

AGE: ______________________

ADDRESS: ______________________

____

S.D.1.

Fig. 9.1. Advertisement used in self-help manual research.

was sent a structured self-help manual containing a step-by-step programme, based on social-learning theory, aimed at reducing drinking.[16] At a follow-up six months later, the group who had received the self-help manual had reduced their drinking more than the control group and reported greater improvements in physical health and alcohol-related problems. The greater reduction in drinking was maintained at a one year follow-up, together with associated benefits.[17]

This was the first study outside the United States to show that a self-help manual was effective in enabling problem drinkers to reduce their drinking; it was the first anywhere to show that such a manual could work in the absence of any personal contact with a counsellor. The problem drinkers in this study had significant alcohol problems, although they were not as seriously disabled as typical alcoholism-clinic attenders. The manual which appears to have helped them contained a cognitive-behavioural self-help programme based upon the principles outlined in Chapters 7 and 8. First, readers were given information about alcohol and its effects, and advised how to decide whether or not alcohol was a problem. They then analysed their reasons for drinking in order to discover the extent to which they were using alcohol as a negative reinforcement, that is, as a way of relieving unpleasant physical or psychological states. The next step was to carry out a 'functional analysis' to identify the internal and external cues associated with heavy drinking. On the basis of this, readers arrived at drinking guidelines regarding where, when, what, how, how much, and with whom they would drink in future. The next stage was concerned with finding alternatives to alcohol, based on the analysis of reasons for drinking. For instance, if drinking served as a recreation, ways of developing other pastimes were discussed; if alcohol was used as a tranquillizer, other forms of anxiety reduction were suggested, and so on. Many other strategies were included which cannot be mentioned here, but the important ingredients were self-help and self-management.

An important subsidiary finding of this study leads us a little further in discussing alternative means of providing advice and help to problem drinkers. As part of the attempt to check on the reliability of the self-reports of drinking by subjects, a sub-sample was telephoned and asked about their drinking, drinking symptoms, physical health, mental state, social relationships, and a number of other areas related to alcohol use. This group did even better than the others, in spite of the fact that they were not given any extra advice by the interviewers, who simply asked factual questions in a friendly and sympathetic manner. It may be, therefore, that this small amount of personal contact boosted the beneficial effects of the self-help manual.

The implications of this tentative finding for providing help to problem drinkers who are unable or unwilling to identify themselves publicly are clear. One can visualize a whole range of telephone, postal, written, audio, or video-taped communications media through which to advise suitable individuals.

Advice can also be delivered in other settings. Jonathan Chick, an Edinburgh psychiatrist, carried out a study in which patients in a general-hospital ward, whose drinking was considered excessive, were given simple written and verbal advice regarding drinking and advised to cut down. Some months after leaving hospital, this group was drinking much less than a control group of similar patients who had received no such advice.[18]

In France,[19] health-screening clinics where people attend for routine medical check-ups have shown very high success rates in persuading excessive drinkers to cut down. Drinkers are given clear factual reasons for reducing drinking (or, in some cases, stopping) where alcohol-related physical damage is identified. These drinkers may well escape the moralistic injunctions which often linger behind similar advice given in clinics devoted to alcohol problems. Not only do these people avoid the stigma of having to declare themselves as having lost control over their alcohol intake, they also escape being 'labelled' as alcoholics. They are simply given sound medical advice to change one of their less healthy habits.

Probably the most successful demonstration of the potential of brief interventions in community settings was conducted in Malmo, Sweden[20]. All male residents of Malmo aged between 45 and 50 were invited to a health screening interview arranged by the Local Department of Preventive Medicine. Of these, 585 men who were in the top 10 per cent of readings of a liver enzyme called gamma-glutamyltransferase (GGT), which is known to reflect the amount of recent alcohol consumption, were selected for study. Of those whose excessive drinking was then confirmed, half were randomly allocated to an intervention consisting of a detailed physical examination and an interview regarding drinking history, alcohol-related problems, and signs of dependence. Subjects were offered appointments with the same physician every three months, plus monthly visits to a nurse who gave feedback about changes in GGT levels. Once drinking had reached moderate levels, frequency of contact was reduced. The other half of the sample were allocated to a control group who were simply informed by letter that they showed evidence of impaired liver function and were advised to cut down drinking.

At follow-up two and four years after initial screening, both groups showed decreases in GGT levels. However, the control group showed a greater increase in the mean number of sick days per individual, more days of hospitalization in the follow-up period, and a strikingly greater number of days in hospital for alcohol-related conditions. At a five year follow-up, the control group showed twice as many deaths, both those probably alcohol-related and deaths of all kinds, as the intervention group. These are the most impressive and well-substantiated results to date in the field of brief interventions, with obvious implications for the widespread application of the method in other countries and with other population groups.

These studies suggest that psychiatric clinics are not the best settings in which to persuade drinkers to change their ways. Of course, there will always be problem drinkers whose complicating psychiatric condition requires psychiatric assistance, but for the bulk of problem drinkers help is probably more effectively located in the community, the work-place, the courts, the general practitioner's surgery, and the general hospital. The general physician who advises a person to cut down drinking because of gastritis is not using a medical model to explain the patient's drinking. On the contrary, the patient is being treated as a rational human being who has a choice over whether or not to drink excessively. This contrasts sharply with some psychiatric approaches, where the whole personality of the drinker is subject to medical investigation as a potentially diseased entity. The demoralizing and disturbing consequences of being seen in this light may contribute to the low attendance rates at traditional alcoholism-treatment centres.

An example of a health-centred approach is being carried out by the Scottish Health Education Group with general practitioners in the Highlands and Islands of Scotland. An aptly named DRAMS kit (Drinking Reasonably and Moderately with Self-control)[21] was devised containing a record card for the GP, a drinking diary, self-monitoring card, and a brief self-help manual adapted from an existing manual written by the present authors. On the record card are a number of signs and symptoms, largely related to physical ailments, which are commonly associated with excessive drinking. If the GP notices any of these signs, or for some other reason suspects that the patient may be drinking excessively, the patient is given a card on which to record alcohol consumption over a two-week period. If the consumption is indeed found to be excessive, the doctor gives the patient the self-help manual. While no evidence for the effectiveness of DRAMS is yet available, it is clear that this procedure provides GPs with a welcome structure through which to do something positive about excessive drinking in the brief consultation period available. While recent government-sponsored reports have recommended that GPs should try to deal with problem drinkers themselves, many simply have not known how to set about doing this. Indeed, it is hard to see how they could have time to provide help along traditional lines.

A rather different method of identifying excessive drinkers among patients of general practitioners was used by Paul Wallace and his colleagues.[22] They gave screening questionnaires, either through the post or the practice receptionist, to the patients of 47 group practices throughout Britain. All those whose replies indicated excessive drinking were invited to a 'life-style and health survey interview' conducted by a trained research nurse. Then, all those whose stated alcohol consumption in the week before interview was over 35 units for men or 20 units for women were invited to take part in a research study and asked to provide a blood sample. Those who agreed were randomly allocated to a 'treatment' group involving an interview from their GP who had been trained in the relevant techniques, another appointment one month later, and up to three subsequent appointments at the discretion of the GP, or to a control group involving no advice from the GP except at the patient's request. The advice given to patients in the treatment group was similar to that described for the DRAMS project and included feedback about the potential harm of the patient's drinking level and practical hints on how to cut down.

The results of this study provide firm evidence of the effectiveness of brief interventions in the general practice setting. At one year follow-up, the proportion of men in the treatment group drinking over recommended limits was reduced from 44 to 26 per cent, while for women these figures were 48 and 29 per cent respectively. The authors claim that, if the results of their study were applied throughout the United Kingdom, general practitioners could be instrumental in reducing to moderate levels the drinking of some 250 000 men and 67 500 women per year. In the light of these calculations, the main conclusion from this research is clear: all general practitioners should be encouraged and trained to deliver brief advice to the excessive drinkers encountered in their practices. Further research is needed to clarify the crucial ingredients of successful brief interventions and to explain why some excessive drinkers fail to respond.

One problem with community-based interventions is that they are usually purely 'opportunistic'—that is, they are directed at persons who have not attended the centre at which the intervention is being offered in order to complain of a drinking problem and may well have to be persuaded that such a problem exists. Thus, the main issue in effecting the intervention is one of motivation rather than the best way to change behaviour once motivation to change can be assumed to exist. The issue of motivation is extremely important at all levels of treatment but becomes especially acute in the types of brief intervention being considered in this section. From this perspective, an important recent development is the elaboration of a set of interviewing principles and techniques designed to increase motivation, known as 'motivational interviewing', by William R. Miller of the University of New Mexico.[23]

Miller argues that the mistake in the past has been to explain poor motivation and lack of compliance in treatment by personality traits and other inadequacies of the client him- or herself. By contrast, when treatment succeeds, it is usually seen as being due to the quality of the treatment programme or the skill of the therapist. Rather, says Miller, motivational problems must be seen in the context of a negotiation between client and therapist. Miller uses well-established principles from social psychology to show how the traditional confrontational approach to problem drinkers is almost designed to lead to the traditional problems of 'denial', 'resistance', and low client self-esteem, not to mention disillusionment on the part of the therapist. As an alternative, Miller de-emphasizes the labelling of 'alcoholic' or 'problem drinker' and instead places heavy emphasis on individual responsibility and internal, rather than external, attribution of change. A deliberate attempt is made to create 'cognitive dissonance' (see p. 265) in the client by subtly contrasting the current problem behaviour with an awareness of its negative consequences and with the client's cherished notions of himself as a person. The therapeutic principle of 'empathic listening', combined with social-psychological principles of attitudinal change and the feedback of the results of objective assessment, are used to channel dissonance towards the desired level of motivation for change. Based as it is in a body of sound psychological findings and a great deal of practical experience and insight, motivational interviewing is proving highly attractive to many workers in the alcohol problems field.

Miller's approach to increasing motivation fits well with a more general framework for thinking about change in the addictive behaviours which has been described by two American psychologists, Jim Prochaska and Carlos DiClemente and which is becoming rapidly influential (see Fig. 9.2). The model proposes that change—say, reducing alcohol consumption to a safer level—can be seen as involving four stages:

(1) a Pre-contemplation stage in which the individual is not aware of any problems connected with drinking and has no desire to change;

(2) a Contemplation stage, in which there is some awareness of problems and the individual may be seen as weighing up the respective advantages and disadvantages of cutting down drinking as against continuing at the present level;

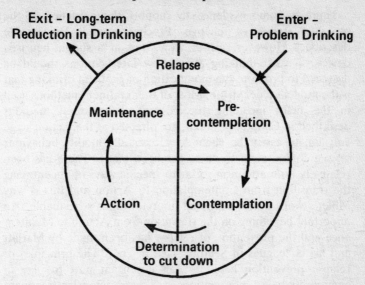

Fig. 9.2. A model of the change process for a problem drinker aiming at reducing consumption [adapted from Prochaska, J. O. and DiClemente, C. C. (1986). Toward a comprehensive model of change. In *Treating addictive behaviours: processes of change* (ed. W. R. Miller and N. Heather). Plenum Press, New York].

(3) an Action stage in which practical efforts are made to reduce drinking;

(4) a Maintenance stage in which the individual is engaged in an attempt to retain the gains that have been made.

In the event that maintenance is unsuccessful, relapse occurs and the person returns to the Pre-contemplation or Contemplation stages; indeed, the history of many drinking problems demonstrates this 'revolving door' pattern which is shown in Fig. 9.2.

There is some evidence to support the existence of the separate stages of change Prochaska and DiClemente describe.[24] However, in our view, it is as a simple heuristic device to assist thinking about how interventions should be designed to promote the modification of problem drinking that is the model's chief virtue. Most of the treatment methods used in the field, including the cognitive-behavioural methods described in Chapters 7 and 8, are directed at the Action stage and aim to assist the client to successfully modify behaviour once a commitment to do so has been made. There has been relatively little attention so far to specific ways of encouraging the transition from Contemplation to Action and this is why Miller's work on motivational interviewing is so valuable. An important beginning on the transition from Action to Maintenance and the prevention of relapse has been made by Marlatt and his colleagues in Seattle[25] (see p. 264). The principles of relapse prevention have just an important part to play in community-based interventions as in more intensive treatment in clinics.

In summary, the social-learning model provides a framework which allows us to intervene with a far wider range of problem drinkers than was previously possible and to make use of many different means of communication which break away from existing therapist-client relationships. This essentially educational format of intervention permits a theoretical synthesis of treatment and certan aspects of prevention, particularly in the area known as 'secondary prevention' (see p. 157). It also allows a far wider range of professionals and counsellors to intervene in an effective way, ranging from general-hospital physicians and nurses, through general practitioners, to company doctors, nurses, welfare officers, health visitors, and several others. We shall turn to the policy implications of this in the Epilogue.

Prevention

That a report on alcohol policies written on behalf of the British government should become available to citizens of this country only because it was published by a Finnish sociologist in Sweden says a lot about the prospects for an effective prevention policy in the UK! The report, *Alcohol policies*,[26] was produced by the Central Policy Review Staff, known as the 'Think Tank', and completed in May 1979 at about the time the new Conservative government was elected. It must have been one of the first tasks of the new administration to suppress the report, which has never been officially published in this country. Running the risk of prosecution under the Official Secrets Act, a number of individuals ensured that it was not buried in some obscure corner of Whitehall by sending copies to Professor Kettil Bruun of the Sociological Institution in Stockholm.

The report made many recommendations, but the central one was that the amount of alcohol drunk per head of population in Britain should cease to rise. This proposal was made because of the abundant evidence showing that the more alcohol a nation consumes, the more alcohol problems of every kind will be found in it (see pp. 127–32).

Whatever the reasons for the suppression of the Think Tank report, it may be no coincidence that the drink industry is fiercely opposed to any government intervention which would halt the rise in alcohol consumption. Neither is it a coincidence, perhaps, that no less than seventy-four Members of Parliament have a direct or indirect financial interest in the alcohol-production industry, according to Mike Daube, formerly of the University of Edinburgh.

So, the government is not keen to upset an industry which produces around two and a half billion pounds each year for the Exchequer (based on figures for 1977). Its official response to the Think Tank report, *Drinking sensibly*, reflected this caution when it rejected indexing of the price of alcohol.[27] This is hardly surprising and, moreover, is a position which cannot possibly lose votes since most drinkers are unwilling to pay more for their beer, whisky, and wine in order to reduce the incidence of what are, to most of them, hypothetical and abstract alcohol problems.

Nevertheless, until alcohol consumption does cease to rise, those attempting to treat the problem drinker, or trying to prevent alcohol problems by other means, are simply wasting their time; numbers will continue to increase in spite of their most valiant efforts. Evidence for this comes from many sources, but one recent study, from Edinburgh University, took advantage of a situation where the real price of alcohol did increase.[28] In March 1981, there occurred a rise in the excise duty on alcoholic drinks which was greater than the increase in disposable incomes and in the retail price index. In other words, alcohol became more expensive in real terms. Comparing the reported drinking of a sample of regular drinkers in 1981-2 and 1978-9, researchers showed an 18 per cent reduction in consumption and an associated 16 per cent drop in the reported adverse effects of drinking, such as having blackouts, getting into fights, being involved in accidents etc. The reduction in consumption was as great for the heaviest drinkers as for the lightest, arguing against the often-held position that the heaviest drinkers will continue to secure their supply of alcohol at any cost to themselves and their families.

The Think Tank report contained a number of other recommendations, relating to licensing laws, alcohol and work, drinking and driving, advertising, treatment of problem drinkers, and many other issues. The theme underlying all these recommendations was that there should be a coherent government policy towards alcohol problems, underpinned by the aim of arresting the increase in per capita consumption. The suppression of the report gives the clearest possible indication that the present government is reluctant to introduce measures which would significantly prevent alcohol-related problems.

Ironically, the government's economic policies seem to have curbed alcohol consumption between 1979 and 1983, though this is likely to be a temporary effect and we can expect further increases if the economy improves. Stopping the rise in per capita consumption will not, however, be sufficient for the development of a maximally effective prevention policy. Education, legislation, and licensing all have parts to play, and it is unlikely that any one preventive measure will be successful in the long term. The policital, economic, and social issues surrounding these measures are hugely complex and outside the scope of this book, but the Think Tank report provides an excellent summary. The most we can do here is to make some comments, from the perspective of a social-psychological model of alcohol use and abuse, about the more salient aspects of prevention.

Fiscal policies

The fact that drinkers of all kinds drink less when alcohol costs them more is quite understandable within the framework of social-learning theory. As we saw in Chapter 4, even chronic alcoholics will reduce, increase, or even stop drinking in a laboratory setting depending on the consequences drinking has for them.

In the natural environment, one of the most immediate, consistent (and thus potent) consequences of drinking is the shortage of money which follows from spending too much on alcohol. Thus, it is not surprising that people tend to drink less as the price of alcohol increases. This effect is not uniform or totally predictable and there will always be some drinkers who are relatively immune to price increases, including the very wealthy. Nevertheless, as the Edinburgh study suggests, even heavy drinkers reduce their consumption as the real price increases.

There are those who think that using price to control the public's drinking is a distasteful infringement of personal liberties and an affront to the assumption that people can make informed choices about their actions. However, to take this to its logical conclusion would lead to a return to something approaching the 'gin epidemic' of the eighteenth century. There can be little doubt that, if alcohol were made much cheaper and much more available, most people would increase their drinking considerably. If that were to happen, there can be equally little doubt that alcohol problems of every kind would mushroom. Thus, if one accepts the principle of *some* regulation, fiscal or otherwise, on the distribution of alcohol, one cannot at the same time employ arguments of principle to object to the use of specific taxation policies. Opponents of fiscal policies must then turn to the more practical implications of using price to control consumption. One such concern has to do with the unfairness of a fixed percentage increase in the cost of alcohol, in so far as it is likely to hit poorer people more than the rich. Other issues relate to the possible expansion of home wine- and beer-making, illegal distilling, and the growth in consumption of imported duty-free alcohol brought in by day-trippers and package holiday-makers. The possibility that families of problem drinkers might suffer material deprivation as a result of alcohol price increases has also been raised, but the findings mentioned earlier, showing that the heaviest drinkers reduced their drinking as much as lighter drinkers following price increases, argues against this point.

Licensing of liquor

British governments have found it necessary to restrict the distribution and availability of alcohol to the public for the last six centuries. The most draconian change in recent times came with Lloyd George's Defence of the Realm Act in 1916, when the opening hours of licensed premises were considerably reduced. This was said to have been a reaction against a loss of productivity in munitions factories caused by drinking which was handicapping the war effort, though the real reasons were more complex.

Changes have also taken place in more recent times. The 1976 Licensing (Scotland) Act considerably relaxed licensing hours, allowing pubs to open till 11 p.m. instead of 10 p.m. and making provision for opening at other times for specific purposes. These latter provisions have created a loophole in the law so that all-day opening is now commonplace. The licensing laws of England and Wales were altered in the 1988 Licensing Act to allow all-day licenses, along with other changes. One as yet unpublished research project carried out in Birmingham by the Court Alcohol Service suggested that introduction of these laws was associated with an increase of approximately 23.6 per cent in magistrates court appearances for alcohol-related offences during the first twenty days of the implementation of the act.

What has social-learning theory to say about licensing laws? First, a balance must be struck between making drinking a 'forbidden fruit' through over-restrictive measures and peppering the environment with drinking cues. The 'swills' of past years, at six o'clock and ten o'clock respectively in Australia and Scotland, testify to the potentially harmful effects of over-restrictive laws. When drinking habits are shaped by laws in this way, modelling ensures that whole generations of drinkers learn to consume alcohol in a manner which is detrimental both to themselves and to law and order. While these licensing laws may make sure that most people do not drink large quantities each week, they also make it inevitable that what is consumed is funnelled into short periods of time. The problems stemming from this type of drinking—drunkenness, violence, public disorder, certain types of accidents—are quite different from those arising in countries where controls on drinking are lax (for example, the wine-producing countries such as Italy, France, Spain, and Germany). In these countries, much more is drunk but is not channelled by licensing laws into relatively brief periods of the day. Thus, the salient alcohol problems in these countries are to do with liver cirrhosis and other danger-ous medical conditions, industrial accidents, automobile acci-dents, and so forth.

To take the other side of the argument, strict licensing laws do restrict the 'drinking cues' that surround drinkers in their environments. If no pubs are open in the afternoon, the drinker is less likely to experience the urge to drink. Witness the relative lack of 'craving' shown by drying-out problem drinkers in hospital and prison; because they know no drink is available and because drinking cues are absent, they experience much less desire to drink than their tremulous condition would lead one to expect. Alcohol is a potent and readily available reinforcer and tranquillizer. The more available it is, the more likely it is that it will be used in preference to less immediate, but also less harmful, means of reducing stress or seeking excitement. For example, whereas in the 1950s a jaded young mother at home with a crying baby may well have coped by having a cup of tea with a neighbour, in the 1980s the same young mother may well take a drink with her neighbour, for no other reason than that it is there and is a more effective way of reducing stress than tea and chat, in the short term at least.

Social-learning theory would thus support restricting the availability of alcohol through licensing, within certain limits. The presence of alcohol on supermarket shelves beside baked beans and bin-liners is almost certain to lead to some people buying a bottle or can on impulse; if they had to go to an off-licence, they might not have bothered. Drinking increases in both individuals and societies, not by clearly thought-out decisions about priorities, but by a series of insidious and often half-conscious, cumulative 'mini-decisions' which gradually build up to well-ingrained habits of purchasing and consumption. Those who sell alcohol make full use of cues to prime these 'mini-decisions', and their advertisements try to enhance and broaden its expected consequences and reinforcements.

One of the arguments put forward by those opposed to strict licensing laws is that alcohol should be 'integrated' with other activities, like eating and leisure pursuits. While there is much to be said for this approach, the Finnish experience must be borne in mind. In the late 1960s and early 1970s, the Finnish authorities sought to improve the drinking habits of the populace, which consisted more or less of drinking to oblivion on one or two nights a week and abstaining on the rest, through a campaign to encourage people to drink wine with meals and, in general, adopt more 'European' drinking habits. There followed a liberalization of the previously stringent laws governing the supply of alcohol. What happened was that the Finns did adopt the European habits but also kept the Finnish! They acquired the taste for 'civilized', moderate drinking, like having wine with meals, but retained the custom of drinking to oblivion. From a social-learning perspective, there was no reason to expect otherwise; habits tend to be relatively independent and the adoption of a new habit is unlikely to supplant an old one unless the cues associated with both are identical, or unless specific attempts are made to suppress the old habit.

Nevertheless, in Scotland at least, drinking has tended until relatively recently to take place in stark, forbidding rooms whose cues positively scream the message, 'Drink, drink!' The attitudes of some publicans tended to reinforce this message; one forbidding landlady known to the authors was once heard to remark to one of her customers, 'You haven't come here to enjoy yourself, you've come here to drink!' Seen in this light, some of the recommendations of the Clayson Report, a Scottish report on licensing laws[29] which argued for changing the drinking environment, can only be welcomed, and it is unfortunate that the government chose to implement mainly those recommendations relating to the extension of licensing hours. If cues for activities other than drinking can be introduced into pubs (say, for eating or games) these may compete with drinking cues and reduce the amount consumed.

Effective policing to ensure that licensees keep within the law is also likely to reduce alcohol consumption and problems. A study carried out by Brian Jeffs, a senior police officer in Devon, in conjunction with Bill Saunders, then of the Alcohol Studies Centre, Paisley, showed that when police made an effort to enforce existing laws, such as those forbidding publicans to sell alcohol to a drunk person, the incidence of drunkenness-related offences fell significantly.[30] This is again in accord with social-learning explanations in which drinking is as much a response to environmental sanctions and rewards as the result of deliberate decisions and choices.

Drinking age

The same balance between restriction and permissiveness must be struck with regard to drinking age. Youngsters brought up by parents who disapprove of their drinking are more likely to have problems once they do finally begin to drink.[31] There is no question that, in adolescence, alcohol can become a weapon with which to rebel against parental authority. Given that most teenagers have drunk alcohol by the age of 14, it is probably better for them to learn to use it at home with parents than on the streets with friends, where they may pick up undesirable habits like drinking rapidly and purely for the dramatic, intoxicating effects. On the other hand, reducing the age at which teenagers are allowed to drink in pubs would serve no good purpose, since the lower the legal drinking age the earlier the young people tend to take their first drink. In fact, the present laws regarding under-age drinking are so widely broken that the Think Tank report recommended they should be much more strictly enforced.

Alcohol and employment

Such is the power of alcohol to entice people to drink and so widespread are the opportunities that, in the absence of sanctions against its use, many people drink in circumstances where it causes a danger to themselves and others. Drinking while driving is one example, drinking at work another. Many industrial accidents are attributable to alcohol.

It is no coincidence that people whose occupations allow them to structure their working day and to work unsupervised have higher rates of alcohol problems than more closely supervised occupational groups. This is especially true where jobs are stressful, where there is separation from friends and family, and where alcohol is readily available.[32] Publicans and hoteliers are at extreme risk of developing alcohol problems because of lack of supervision and high availability, among other reasons.

As mentioned earlier, the worst thing the colleagues of a problem drinker can do is to cover up for lateness, inefficiencies, and botch-ups at work. In doing so, they are protecting the drinker from the natural consequences of such behaviour and are thus prolonging its damaging effects. Unfortunately, people do not usually learn by merely being told to drink less, but from the hard experience of last warnings, financial penalties, and even dismissals.

If this applies to the problem drinker, then it applies to all drinkers, assuming that the social-learning view of a common explanatory model is accepted. If you have a job where drinking at lunchtime or even during working hours is tolerated, you will be more likely to drink during the day and, in any group of employees, a few will start to drink too much precisely because of this laxity of sanctions. As a consequence, some will learn the dangerous habit of relief drinking (see p. 210).

Given that a large part of our lives is spent there, sanctions against drinking at work must be one of the most potentially effective preventive strategies which could be devised. If all workplaces were devoid of cues for drinking, a major step towards prevention would be taken. In Canada and France, government policies have been introduced to try to keep alcohol consumption as far away from the workplace as possible, though the UK government has not been as active.[33]

An additional benefit of such a strategy is that problem drinkers become conspicuous before their habit becomes too well-developed. As an alternative to the disciplinary procedures which would in any case have followed their misdemeanours, they can be offered advice on how to cut down drinking. The potency of the underlying sanctions, and the relative weakness of the drinking habit due to its early detection, ensure that industrial programmes of this type can be very effective.[34]

Drinking and driving

One in five road accident deaths is caused by excessive alcohol consumption, amounting to 1200 deaths per year. When the breathalyser was first introduced in Britain, there was a dramatic decline in road accident deaths and it is estimated that, in the first seven years of its operation, some 5000 lives were saved. Since then, fatalities have steadily risen. While this partly reflects the increase in national alcohol consumption over this period, it is likely that the public has learned that the risks of being breathalysed by the police are low. Though the consequences of being convicted may be severe for some people, say those whose livelihood depends upon a clean driving licence, if they do not perceive the risks of conviction to be high, its deterrent value will be greatly diminished.

Successive governments have recognized these facts, yet none have found parliamentary time to implement the recommendations of the Blennerhassett Committee, a government-sponsored body which made a number of recommendations for changing the drinking-and-driving laws.[35] One of the most controversial recommendations was that the police should be allowed to breathalyse drivers at their discretion—the so-called random test recommendation. While many view this as a further escalation of the already over-developed powers of the police, from a social-learning point of view discretionary testing would have some deterrent effect on potential drunk drivers. This is partly because random testing would diminish the sense of control of the drunk driver over whether or not he is stopped by the police. At the moment, drunk drivers can nurture illusions about their capabilities for driving while intoxicated, thus psychologically avoiding both moral qualms and practical fears of conviction. The introduction of random testing, as in New South Wales in Australia, would strike a major blow against these illusions.

Recent legislation in Britain provides a welcome tightening of one aspect of the drunk-driving laws. 'Serious drink drivers' will not get their licenses back after disqualification until they have passed a medical examination. To qualify as 'serious', they must have had a blood alcohol level of more than 200mg/100ml, have refused to provide a breath, blood, or urine sample, or have been disqualified for two drink-driving offences in ten years.

Advertising

The issues surrounding the advertising of alcohol are complex and we will not attempt to cover them all here. While some people have called for a ban on alcohol advertising, there is no available evidence to show that this would influence alcohol consumption. Nevertheless, there are intuitive grounds for suspecting that the advertising of alcohol aimed at groups whose drinking is not up to its full potential may influence how much they drink. For instance, surveys have shown that, in the last few years, young, single, employed women in their twenties have increased their consumption of alcohol more than any other group.[36] The wave of advertising aimed at young women which hints strongly at adventure, sexual and otherwise, through drinking certain vodkas and vermouths has at least followed, and perhaps contributed to, these trends.

In Chapters 6 and 8 we outlined the concept of modelling and showed how behaviour can be influenced by simply watching other people do things—whether in real life or on film. If children's aggressiveness can be increased by watching films and cartoons depicting violence and aggression,[37] it would be fair to assume that drinking behaviour can be similarly influenced. Of course, it is not only in advertisements that the use of alcohol is modelled; films, television, videos, books, and magazines all communicate potent messages about what effects alcohol has, when it is drunk, who drinks it, for what reasons, and so on.

Research has shown that modelling will more readily take place when the model is seen to be rewarded for doing something. The symbolic values of drinking—sexual adventure, status, excitement, popularity—are powerful rewards which advertisers continue to portray their models as receiving. This is despite codes of practice, set up by the Advertising Standards Authority and by the Independent Broadcasting Authority, restricting the activities with which the use of alcohol may be associated. The Think Tank report recommended that the government should try to ensure that the main aim of alcohol advertising is to influence the choice of beverage consumed, not how people make that choice—alcohol should not be portrayed as having relaxing or other rewarding effects. Instead, advertising should concentrate on those merits of an individual product which distinguish it from its rivals. This is, in fact, what the alcohol industry professes to be doing; it denies that advertising does more than influence the choice of beverage.

The most cursory glance at any drinks marketing publication reveals that this last assertion is mere cant. To take one example, women are widely discussed as being an 'underdeveloped market' for alcohol in the advertising and marketing press. The results of this can be seen in the advertising directed at young women, associating drink with feminine independence and sexual adventure. The advertisers are not trying to create 'brand switching' when they talk of this promising market: they are attempting to persuade more women to drink and those who drink already, to drink more.

Britain is also regarded as an 'underdeveloped market' by these same marketing specialists because national per capita consumption in Britain is much lower than in most of its European neighbours. Again, it is simply nonsense to assert that the drinks industry is simply wanting to change what people in Britain drink—they are wanting the British to drink more alcohol of every kind.

In the last two years the British government was at last shamed into doing something about alcohol problems. One of the main causes of this was not, however, the death, injury, and illness caused by alcohol, but rather the embarrassment at the drunken violence shown by English football fans at home and abroad. The banning of English teams from European competitions in the aftermath of the Heysel Stadium disaster in Belgium was a major prompt to the government setting up an interdepartmental committee on alcohol chaired by Mr John Wakeham.

In the eyes of some observers, the aggressive and often drunken behaviour of some English supporters was abetted by the contents of drinks advertisements in which consumers are encouraged to choose drinks on the basis of their alcoholic strength and engage in loud, macho group activity. This bore uncanny resemblance to the behaviour of the so-called 'lager louts' on their violent sprees.

Only after this reached intolerable levels was the Wakeham committee spurred into action in relation to advertising codes issued by the Independent Broadcasting Authority and the

British Code of Advertising Practice. These codes include such strictures as: 'Advertising should not give the impression of being inducements to drink because of its higher alcohol content or intoxicating effect . . .'. Examine the two following, recently-used adverts in this light: 'Spend some time out of your tree . . .'—Woodpecker cider; 'Some lions prefer a bigger bite . . .'—Kestrel super-lager.

Aitken and his colleagues at Strathclyde University in Glasgow failed, however, to detect any change in the contents of drinks advertisements after these new regulations were introduced.[38] They also showed that drinks advertising appeals to 13- to 18-year olds. Indeed, one advert for Carling Black Label lager was found to be the most popular commercial of all among 13-year olds. Young drinkers under 25 are the lager and cider market's biggest customers, and thus it should not be surprising to find youth cult figures such as Jonathon Ross appearing in adverts (again in contravention of the IBA code: 'No liquor advertisement may feature any personality whose example young people are likely to follow'). Codes against associating alcohol with masculinity, physical prowess, and sexuality are also regularly broken as Strongbow arrows thud into doors, boring parties are overrun by Tenants gatecrashers, and vodka lurks in the misty light of sexual promise.

Young people from 15 to 25 have the greatest amount of leisure time and free disposable income. A large part of that income is spent on alcohol, and youth culture increasingly centres around alcohol. Thus, those in this age group are the drinks industry's best customers. And given that just 30 per cent of all the alcohol is drunk by just 3 per cent of the population, heavy drinking cannot be discouraged by the drinks industry and, indeed, must be encouraged because otherwise profits would slump.

Heavy drinking is fostered by associating alcohol with sexuality, sexual identity, popularity, and social success. These are key concerns of adolescents and young adults. While maleness has been linked with drinking prowess for many hundreds of years, the association of female sexuality with drinking is very recent and may well be a dangerous creation of the advertising industry. The fact that, over thousands of years, drinking has not been associated with female sexuality may be in part due to the awareness in earlier cultures of the dangers of drinking by pregnant women.

Education

Many thousands of pounds were spent in Britain during the 1970s on the 'clunk-clink' advertising campaign aimed at encouraging car drivers and passengers to wear seat-belts. The effects were negligible, and only when the wearing of seat-belts was made compulsory through an Act of Parliament in 1983 was there an increase in seat-belt use and a consequent saving of many hundreds of lives each year, not to mention the avoidance of tens of thousands of crippling and disfiguring injuries. Those who argue that education is the only answer to the prevention of alcohol problems would do well to bear the seat-belt experience in mind. This includes Her Majesty's Government, which in the document *Drinking sensibly*, rejected fiscal policies in favour of educational measures.[27]

That education should have so little effect in these circumstances is understandable from a social-learning viewpoint. Most humans are sustained in their paths towards distant or intangible goals only through the correcting and encouraging mediation of short- and medium-term rewards and punishments. All but the most saintly need some nods of encouragement or some tangible rewards to keep them at their task. When education aims to alert people to a course of action containing intrinsic reinforcements, one can be more optimistic about its chances of success. Thus, education may have had some effect in discouraging smoking because the benefits of stopping smoking are quickly apparent to many who do so.

Similarly, Ministry of Agriculture education about potato blight or Colorado beetle is likely to influence farmers and small-holders who will reap tangible benefits from obeying this advice.

In cases where education is aimed at influencing particular items of behaviour, where the perceived benefits are psychologically, physically, temporarily, and/or statistically distant (such as death or injury in a road accident), it is unlikely to be successful, especially with the 'risk-takers' of our community. Prior to 1983, the pay-off to an individual for starting to wear a seat-belt was that he or she would be less likely to be killed or seriously injured in a crash. Given that most people, most of the time, do not really believe such things could happen to them, this pay-off was simply not enough. Once legislation was introduced, the pay-off for wearing a seat-belt became the avoidance of a fine; this was tangible and enough of an incentive to cause seat-belt use to rocket at a stroke.

The relevance of this to alcohol is obvious and there are many parallels between advice about seat-belt use and advice about drinking. Both are based on danger in the distant future which will only strike a small proportion of those hearing the message; thus the dangers can easily be perceived as 'unlikely to happen to me'. In many ways, raising the price of alcohol, reducing the public's tolerance of alcohol use on the roads and at the workplace, and decreasing public tolerance of drunkenness in general are measures roughly equivalent to making seat-belt use compulsory. They are all means of introducing short-term incentives to encourage adherence to a long-term goal.

So is alcohol education a waste of time? Yes, if carried out in the absence of tangible, short-term incentives for the reduction in drinking which is being advised. If, on the other hand, it *is* closely linked with such concrete measures, not only is it potentially useful but indispensable. To return to the case of seat-belts, it would have been politically impossible to introduce legislation if the public had not been educated about reasons for wearing them. Similarly, the public would have rebelled against the progressive raising of the cost of cigarettes if they had not been well informed about the links between smoking, lung cancer, and many other diseases.

A primary task of alcohol education should therefore be to acquaint the public with the practical reasons for drinking moderately. This is easier said than done, because there is disagreement, even among specialists in the field, about what the content of these messages should be and, in particular, where safe levels of alcohol consumption should be located. Assuming these difficulties can be overcome, and that the public becomes more aware of the effects of heavy drinking, education should then focus on putting over the case for tighter controls on alcohol use at work, on the roads, and in the pubs, and for a linking of the cost of alcohol to the retail price index.

The role of education with the young presents greater complexities. There is hardly any evidence to suggest that alcohol education in schools has had any measurable effects on subsequent drinking. Yet it might be argued that education should not aim to change behaviour. Schools and colleges are institutions where young people could learn basic facts about alcohol and its use and where, through personal contact with teachers, they could discuss their own use of alcohol. But most young people are more influenced by peer groups than by teachers, which means that even if they develop favourable attitudes to moderate alcohol use in school, the norm of the peer group could rapidly overturn these attitudes. School programmes can only work in the context of a wider change in public attitudes to alcohol, and these changes will take place over decades rather than years, if they happen at all. Currently, the most promising educational approaches are secondary-prevention programmes, in which education is aimed at those for whom drinking is causing some tangible and current problems. These groups (young offenders, attenders at health screening clinics, hospital patients with alcohol-related physical disorders, etc.) are likely to respond to education precisely because they can see some immediate, concrete reason for following the advice given.

Conclusions

This sketchy discussion of the implications of a social-learning theory account of alcohol use for treatment and prevention cannot be a substitute for a discussion of the variety of issues relating to political, economic, and social factors in the response to alcohol problems. As psychologists, our focus has been the individual problem drinker, but this does not mean that individual treatment should be the main element of the response. On the contrary, individual treatment in the absence of effective prevention policies is a vain enterprise. In the Epilogue we shall outline some of the broader issues surrounding alcohol policy.

We should also make clear that the new paradigm of problem drinking outlined in the last four chapters must be seen as only a first step towards placing our understanding of alcohol problems on a sounder scientific basis. Social-learning theory does not yet provide a tightly knit body of knowledge from which to draw precise deductions about the nature of problem drinking and the best way of responding to it. Rather, it offers an alternative language or set of concepts for understanding drinking and drinking problems, together with some empirical backing. The paradigm is patchy and lacking sufficient rigour in many places but, as we trust the preceeding pages have demonstrated, it has great potential and is already more useful than that which it aspires to replace. For the present at least, it is the best bet for advancing towards a comprehensive theory and effective practical response to problem drinking.

Epilogue: the way ahead

Alcohol gives pleasure to tens of millions of men and women in Britain. It provides employment, directly or indirectly, for hundreds of thousands; it inspires song and verse; it symbolizes celebration, companionship, and escape from the drudgery of everyday life. In many circumstances and if used with discretion, alcohol is good for most people.

At the same time, the excessive use of alcohol brings misery to literally millions of problem drinkers and their families in this country; suicide, fatal illness, violence, problem children, depression, anxiety, and a score of other serious disorders can be attributed in varying degrees to excessive drinking.

As a society, what price are we prepared to pay for the undoubted pleasures alcohol gives us? Prohibitionists argue that no pleasure can justify the misery and that we should seek our entertainment from other sources. Conversely, supporters of the lobby trying to prevent the introduction of curbs on alcohol consumption may seek other causes than alcohol for the disorders mentioned above. Both views are unreasonable. One has only to look at the disastrous effects of prohibition in such countries as the United States and Finland in the 1920s to see how important alcohol is to people and the lengths they will go to get it. The evidence outlined in Chapter 4 (pp. 127–32), on the other hand, clearly shows that the incidence of alcohol problems is closely related to how much a nation drinks.

The ambivalence over the use of alcohol which exists in many countries, especially in Northern European countries, reflects this 'Jekyll-and-Hyde' picture of the drug ethyl alcohol, and has itself produced many problems. In the north of Scotland, for instance, roughly twice as many people feel ashamed about their drinking as in the south-east of England, even though they drink on average much the same.[1] When shame and guilt become attached to any behaviour, the chances of it becoming a problem will increase dramatically, for all the reasons outlined in Chapters 6, 7, and 8.

How, then, does one find a way through this maze of conflicting opinion? A first step is to declare unreservedly that moderate drinking is a great pleasure. Too often, experts in the field are loth to admit this publicly. We must not fall into the same trap as the 'neo-prohibitionists' who, while realizing that the demon drink will never be exorcised from the world, nevertheless purse their lips and frown disapprovingly about it. Too many of these pundits purse their lips on the lecture platform and then open them gladly to the pint of beer in the pub afterwards! Such double standards help the cause of the problem drinker not one bit.

While accepting the pleasures of moderate drinking, however, we must also accept that the problems of alcohol have been under-recognized, in the same way as the dangers of smoking were not well known twenty years ago. The blame for this must lie partly at the door of protagonists of the disease model who have induced a false sense of security among drinkers by perpetuating the myth that some mysterious biological entity separates alcoholics from the rest of drinking humanity. So pervasive is this idea that many problem drinkers fend off doubts about their drinking by choosing to believe that they don't have 'it'. Interrogations surrounding the question, 'Are you an alcoholic?' have been as arcane and vitriolic as those concerning the number of angels on the head of a needle; and they have had a similarly retarding effect on logical and scientific advancement.

A second step should therefore be to bring problem drinking back into the realm of 'normal' drinking. Cars are wonderful inventions, yet car accidents are horrific. Drinking moderately is a fine activity, yet problem drinking is tragic. Most drivers accept some responsibility for road-safety, but most drinkers do not yet accept responsibility for drinking-safety. This is what has to change; and in the same way that there is no sanctimonious prohibitionism lurking beneath the debates on road-safety, so we must get rid of the guilt- and conflict-arousing sentiments which are applied to drinking. If one is genuinely concerned about problem drinking, one must accept personal responsibility for supporting the moves that constitute the only real way in which drinking problems will be significantly reduced—a stabilizing of and ultimately a reduction in the total amount of alcohol consumed. Despite the complications surrounding this issue mentioned in Chapter 9, this can only be achieved by gradually increasing the cost of alcohol, at least in line with inflation—a measure to which the government has so far refused to commit itself. (A pressure group called Action on Alcohol Abuse devoted, among other things, to changing government policy in this respect is active in the UK.[2])

Another issue relating to reduction in consumption stems from the fact that a small proportion of drinkers consume a large proportion of the alcohol drunk in our society. One study found that 3 per cent of drinkers drink 30 per cent of the alcohol.[3] If the drink industry is serious about wanting to reduce the incidence of problem drinking, it must be prepared to accept big reductions in consumption. For the way things stand at present, the industry depends for much of its profits on the heaviest drinkers, from whom most problem drinkers will emerge. For this reason, it is unrealistic to expect voluntary measures from the drink industry which will affect the incidence of problem drinking in any noticeable way; and this is why the government must act. Only when seat-belt legislation was brought in did road-safety improve dramatically, and a similar move is necessary with respect to alcohol.

The British government's Wakeham Committee which was mentioned above (p. 319) has, however, set its face firmly against the recommendations of the 'Think Tank' (see p. 307).

It has confined itself firmly to the problems of public order associated with alcohol and is ignoring the central issues of price and availability. An all-too-frequent recipe for alcohol problems issued by the Wakeham committee is 'education'; as mentioned in the last chapter, education may be ineffective in the absence of accompanying encouragements and sanctions at the practical level.

If education is to be an effective arm of prevention, which so far has not been the case, it must turn its focus as much towards the social as the individual level. Government policies are likely to be the major tool for limiting alcohol problems, and therefore educators must address themselves to these measures and educate the public as to the need for them. To restrict themselves to a continual 'nagging' about drinking less is rather unfair to the individual because, as we saw in Chapters 6, 7, and 8, individuals are to a considerable extent at the mercy of environmental and social forces in determining how much they drink. Yes, the individual has a choice and must be well-informed so as to be able to make that choice, but the individual focus must not be allowed to detract from education regarding the social, economic, and political forces which mould drinking so powerfully.

One factor which may render all educational efforts in Britain worthless over the next four years, if it goes ahead in its present form, is the 1992 single market agreement of the EEC. According to present plans, 1992 will see the price of a bottle of spirits in Britain fall by £2.37, wine by 70 p. and a pint of beer by 15 p. This will mean an estimated 28 per cent increase in the amount of alcohol consumed nation-wide. This in turn will result in a comparable increase in deaths, illness, accidents, social problems, and other family disruption. Many people reading this will die younger than they should—of liver cirrhosis, cancer, accident, and so on—because of cheaper alcohol leading them to drink more. Women in particular will suffer because they make up a market which the drinks industry is desperately trying to foster.

Hence the dubious status of 'education' if implemented without attention to other, much more important determinants of how much people drink. Yet the British government refuses to face up to these issues, principally because of the lobbying power of the drinks industry whose voice is very powerful in the governments of most countries. Recently, the alcohol and tobacco industries have formed an alliance as the following quotes from an editorial of *Harpers*, the wine and spirits trade house journal, 18 September 1987, demonstrate:

. . . the time could now be right for a further advance in the trade stance against the anti-alcohol lobby . . . a proposal is being considered . . . to set up regular meetings with the Tobacco Advisory Council . . .

. . . a former Labour Minister has already expressed publicly his own view that any government would be greatly deterred by a joint lobby of the tobacco, drink, and gambling industries and he was urging these industries to be much more vociferous in defending their legitimate interests . . .

If one considers the fact that over 70 Members of Parliament in Britain, including some Government Ministers, may have vested interests in the drinks industry, then the power of the drinks lobby is clear.

A major argument used by those influenced by the drinks industry lobby is that alcoholics will obtain alcohol no matter the price, and the ordinary 'social drinker' should not be penalized for his or her pleasures. This flies in the face of evidence showing that consumption and harm reduces across the board when the real price of alcohol increases, even by a small amount. This happened between 1979 and 1982 in Britain, when alcohol became more expensive in real terms for the first time in thirty years and when falls were observed in several indices of harm, including admissions to hospital for alcohol dependence, convictions for drunkenness, and incidence of liver cirrhosis.

The disease model serves well those who perpetrate the myth of 'the alcoholic will get his drink no matter what', though the fact that 20 per cent of the population drinks 80 per cent of the alcohol suggests that the main customers of the drinks industry are more than light social drinkers. Indeed, some licensed premises would not survive financially without the excessive drinking of a small number of heavy drinking customers. A severely dependent problem drinker on the equivalent of a bottle of spirits per day, who drinks half his or her alcohol in a particular pub or club, would spend more than £4000 per year in that place. Such a person would be part of the 3 per cent of the population who drink just under a third of all the alcohol consumed and, as such, would be a key contributor to the profits of the drinks industry.

It is ironic that, as the disease model of alcohol problems wanes as a credible explanatory tool, so knowledge about the physical consequences of heavy drinking, ranging from testicular shrinkage to brain damage, continues to grow. It is increasingly apparent that the medical profession is coming, in one way, to have an expanding role in alcohol problems and, in another way, a much reduced role. The increased role is likely to be among general practitioners, company doctors, and hospital physicians, who can detect heavy drinking at an early stage. They are thus in a powerful position to influence the drinking behaviour of hitherto unidentified problem drinkers without labelling or stigmatizing their patients and to give adult advice to adults about the physical consequences of their drinking. The evidence is that this can be extremely effective.

The reduced role of the medical profession must surely be in counselling services for more serious problem drinkers. The day of the Alcoholism Treatment Unit with its unnecessary medical paraphernalia is over; community services are what are needed and large numbers of counsellors are required to make any impression on the great number of problem drinkers. (This was recommended in an influential report by the Advisory Committee on Alcoholism to the Department of Health and Social Security in 1978.[4] The report proposed a downgrading of the status of Alcoholism Treatment Units, together with an expansion of services provided by 'front-line' professionals like general practitioners, community nurses and social workers.) Medical expertise is simply not necessary in the bulk of counselling work and, as we hope to have shown, medical models of alcohol problems have been a considerable hindrance to the development of effective services over the last twenty years. There will always be medical problems arising from problem drinking which require medical aid, but treatment of these sequelae should not be confused with treatment of problem drinking itself. Moves in this direction are already being made and, in various parts of Britain, community alcohol teams are now offering a truly comprehensive service to problem drinkers following the closure of Alcoholism Treatment Units. It is likely that similar developments will occur elsewhere.

The way ahead for the response to problem drinking, as we see it, is predicated on the demise of the disease perspective on alcoholism. Assuredly, however, there will be major obstacles along this path. For one thing, it must be candidly stated that some members of the medical profession are ferociously possessive about what they consider to be their own 'turf' and most reluctant to allow anybody to redraw its boundaries, especially if the proposed reorganization conveys any hint of reduced power or status. Hence, a strong objection to the new paradigm can be confidently expected—the familiar cry that abandoning the disease theory means abandoning alcoholics to their fate. We must therefore repeat for the last time that the adoption of the social-learning paradigm does not entail a return to punishment and neglect. On the contrary, it promises more rational and effective means of improving the problem drinker's lot.

Medical opposition to the new paradigm would be doubly unfortunate because of the inescapable reality that it could only be fully effected with the medical profession's active support and co-operation. This could hardly be otherwise in view of that profession's continuing power and influence. We derive considerable encouragement, however, from the fact that some British psychiatrists have been in the vanguard of innovative thinking about alcohol problems, particularly in the development of the community response to problem drinking heralded by the DHSS Advisory Committee's report.[4]

A more recent development is the growth of private clinics for the treatment of alcoholism in Britain, as part of the larger upswing in private medicine of all kinds. This British development springs directly from the tremendous boom in profit-making facilities in the United States, where alcoholism treatment is now a multi-million per annum industry. The successful take-off of such commercial enterprises here is self-evidently reliant, not merely on the retention of the disease perspective, but on an enlargement of its popularity. This will be of benefit to none but a few well-off problem drinkers but the concerted promotion of the disease theory it will entail will impede progress on more social and political fronts.

A final challenge to the new paradigm, in some ways the most regrettable, will almost certainly come from the men and women who are affiliates of AA. The influence of AA, especially its influence on media presentations of problem drinking, is a significant factor in the British situation. Opposition to date has centred on the notion of controlled drinking for problem drinkers but, although it cannot be denied that some implications of the evidence strike at the heart of AA beliefs, this opposition is often based on a misconception of what controlled-drinking exponents are trying to do. Once more, nobody is urging all problem drinkers to abandon abstinence from alcohol or to ignore advice to take it up. Abstinence has an essential role to play in the individual treatment of serious problem drinkers.

There is another, more general way in which the reaction of AA may prove a hindrance to the developing paradigm. A well-known article published in 1976, entitled 'The future of alcohology: craft or science?',[6] discussed the difficulties which had arisen from the influx of scientists and professionals in a field that had, up to then, been dominated by para-professional workers, especially AA members. The clash of a 'scientific' with a 'craft' model had created unique tensions, which could not be easily resolved since the two models were incompatible. Unfortunately, the authors of this article did not answer their own question and the reader was left wondering what solution to the dilemma was being proposed.

We wish to answer the question here. We are well aware of the dangers of 'scientism'—the unthinking replacement of old deities with the new god of science—and by no means wish to suggest that the response to drinking problems should become the exclusive province of psychologists, social workers, or any other professional group. As has been made abundantly clear, volunteers and para-professionals have an indispensable part in the projected plan. But we are convinced that the way ahead will not be found in a continued adherence to folk-lore, myth, and tradition. It must be based instead on an empirically grounded, rational, and coherent national policy for the treatment and prevention of problem drinking.

Notes and references

1. Introduction

1. Some recent general introductions to alcohol problems, all stressing somewhat different aspects of the subject, are as follows: Plant, M. A. (ed.) (1982). *Drinking and problem drinking*. Junction Books, London (a fairly comprehensive overview); Rix, K. and Rix, E. (1983). *Alcohol problems: a guide for nurses and other health professionals*. John Wright, Bristol (a particularly good review of the physical aspects of alcohol use and abuse); Davies, I. and Raistrick, D. (1981). *Dealing with drink*. BBC Publications, London (oriented more towards the practicalities of counselling problem drinkers, but a good general introduction too); Taylor, D. (1981). *Alcohol: reducing the harm*. Office of Health Economics, London (a useful booklet containing some key facts and figures); Grant, M. and Gwinner, P. (eds.) (1979). *Alcoholism in perspective*. Croom Helm, London (a useful and brief overview of most of the important issues in the field); Royal College of Psychiatrists (1979). *Alcohol and alcoholism*. Tavistock, London (highly informative review of most of the important issues); J. and J. Chick (1984). *Drinking problems: advice and information for the individual, family and friends*. Churchill Livingstone, Edinburgh (practical advice and information for problem drinkers and those around them); Robertson, I. and Heather, N. (1986). *Let's drink to your health!* B.P.S. Publications, Leicester (a self-help guide to responsible drinking).
2. Kuhn, T. S. (1970). *The structure of scientific revolutions* (revised edn). University of Chicago Press, Chicago.
3. Jellinek, E. M. (1960). *The disease concept of alcoholism*. Hillhouse Press, New Haven.
4. Cahalan, D. (1970). *Problem drinkers: a national survey*. Jossey-Bass, San Francisco.

2. The historical context

1. Quoted in Longmate, N. (1983). Alcohol and the family in history. In *Alcohol and the family* (ed. J. Orford and J. Harwin). Croom Helm, London. Longmate's splendid history of the Temperance Movement, *The waterdrinkers* (1968), Hamish Hamilton, London, has also been used extensively as a source in this chapter.

2. See Coffey, T. G. (1966). Beer Street: Gin Lane. Some views of 18th century drinking. *Quarterly Journal of Studies on Alcohol*, **27**, 669–92.

3. Adapted from Spring, J. A. and Buss, D. H. (1977). Three centuries of alcohol in the British diet. *Nature*, **270**, 567–72.

4. Quoted in Lender, M. E. and Martin, J. K. (1982). *Drinking in America: a history*. The Free Press, New York. This has also proved an invaluable source of material for this chapter.

5. See Lender, M. (1973). Drunkenness as an offense in early New England: a study of 'Puritan' attitudes. *Quarterly Journal of Studies on Alcohol*, **34**, 353–66.

6. Gusfield, J. R. (1963). *Symbolic crusade*. University of Illinois Press, Urbana, Illinois.

7. See Thorner, I. (1953). Ascetic Protestantism and alcoholism. *Psychiatry*, **16**, 167–76.

8. Bacon, S. D. (1967). The classic Temperance Movement of the USA: impact today on attitudes, action and research. *British Journal of Addiction*, **62**, 5–18.

9. Levine, H. G. (1978). The discovery of addiction: changing conceptions of habitual drunkenness in America. *Journal of Studies on Alcohol*, **39**, 143–74.

10. The *Inquiry* was reprinted in the *Quarterly Journal of Studies on Alcohol* (1943), **4**, 321–41.

11. Also partly reprinted in the *Quarterly Journal of Studies on Alcohol* (1941), **2**, 584–5.

12. Described in Maxwell, M. A. (1950). The Washington Movement. *Quarterly Journal of Studies on Alcohol*, **11**, 410–51.

13. Harrison, B. (1970). *Drink and the Victorians*. Faber, London.

14. See, for example, Trice, H. M. and Roman, P. M. (1970) Delabeling, relabeling and Alcoholics Anonymous. *Social Problems*, **17**, 538–46.

15. See Lender, M. E. (1979). Jellinek's typology of alcoholism: some historical antecedents. *Journal of Studies on Alcohol*, **40**, 361–75.

16. See, especially, Paredes, A. (1976). The history of the concept of alcoholism. In *Alcoholism* (ed. R. Tarter and A. Sugarman). Addison–Wesley, Reading, Massachusetts.

17. The early years of the Alcoholism Movement are described by Keller, M. (1976). Problems with alcohol: an historical perspective. In *Alcohol and alcohol problems* (ed. W. Filstead, J. Rossi, and M. Keller). Ballinger, Cambridge, Massachusetts.

18. Straus, R. (1976). Problem drinking in the perspective of social change. In *Alcohol and alcohol problems* (ed. W. Filstead, J. Rossi, and M. Keller). Ballinger, Cambridge, Massachusetts.

19. Room, R. (1972). Comments on 'The alcohologist's addiction'. *Quarterly Journal of Studies on Alcohol*, **33**, 1049–59.

20. Rubin, J. L. (1979). Shifting perspectives on the Alcoholism Treatment Movement 1940–1955. *Journal of Studies on Alcohol*, **40**, 376–85.

21. Quoted in Orford, J. and Edwards, G. (1977). *Alcoholism*. Oxford University Press.

22. DHSS (1973). *Community services for alcoholics*, Circular 21/73. Department of Health and Social Security.

23. An account of the early history of AA may be found in Lender, M. E. and Martin, J. K. (1982). *Drinking in America: a history*. The Free Press, New York.

24. *Alcoholics Anonymous* (1939). Works Publishing Inc., New York.

25. Jones, R. W. and Heldrich, A. R. (1972). Treatment of alcoholism by physicians in private practice: a national survey. *Quarterly Journal of Studies on Alcohol*, **33**, 117–31.

26. See Robinson, D. (1977). *From drinking to alcoholism: a sociological commentary*, Chap. 5. Wiley, London.

27. See Ogborne, A. C. and Glaser, F. B. (1981). Characteristics of affiliates of Alcoholics Anonymous: a review of the literature. *Journal of Studies on Alcohol*, **42**, 661–75.

28. See Robinson, D. (1979). *Talking out of alcoholism: the self-help process of Alcoholics Anonymous*. Croom Helm, London.

29. Robinson, D., personal communication.

30. From the 1955 edition of the AA 'Big Book'. Alcoholics Anonymous (1955). *The story of how thousands of men and women have recovered from alcoholism*. AA World Services, New York.

31. For a review of the relevant evidence, see Miller, W. R. and

Hester, R. K. (1980). Treating the problem drinker: modern approaches. In *The addictive behaviors* (ed. W. R. Miller). Pergamon Press, New York.

32. Brandsma, J. M., Maultsby, M. C., and Welsh, R. J. (1981). *The outpatient treatment of alcoholism: a review and comparative study*. University Park Press, Baltimore.

33. Lefever, R. (1986). *Royal College of General Practitioners Members reference book*. Royal College of General Practitioners, London.

34. These figures appear in a publication of the Hazelden Foundation. See Jones, K., *et al.* (1986). *Treatment Benchmarks*. Hazelden Foundation, Minneapolis.

3. How many disease concepts of alcoholism are there?

1. Jellinek, E. M. (1960). *The disease concept of alcoholism*. Hillhouse Press, New Haven.

2. Good surveys of the medical complications of excessive drinking may be found in Rix, K. and Rix, E. (1983). *Alcohol problems: a guide for nurses and other health professionals*. John Wright, Bristol; and in the Report of a Special Committee of the Royal College of Psychiatrists (1979). *Alcohol and alcoholism*, Tavistock, London.

3. Alcoholics Anonymous (1939). *Alcoholics Anonymous*. Works Publishing, New York.

4. Williams, R. J. (1948). Alcoholics and metabolism. *Scientific American*, **179**, 50–53.

5. Edwards, G. (1970). The status of alcoholism as a disease. In *Modern trends in drug dependence and alcoholism* (ed. R. V. Phillipson). Butterworth, London. This contains a useful discussion of the evidence for and against earlier disease concepts.

6. Freud, S. (1955). Mourning and melancholia. In *The standard edition of the complete psychological works of Sigmund Freud* (ed. I. Starchey). Hogarth Press, London.

7. Freud, S. (1955). Three contributions to the theory of sex. In *The standard edition of the complete psychological works of Sigmund Freud* (ed. I. Starchey). Hogarth Press, London.

8. Menninger, K. (1938). *Man against himself*. Harcourt Brace, New York.

9. McCord, W., McCord, J., and Gudeman, J. (1960). *Origins of alcoholism*. Stanford University Press, Stanford.

10. McClelland, D. C., Davis, W. N., Kalin, R., and Wanner, E. (1972). *The drinking man*. The Free Press, New York.

11. Jellinek, E. M. (1952). Phases of alcohol addiction. *Quarterly Journal of Studies on Alcohol*, **13**, 673–84.

12. Figure 2 shows a later, revised version of the original chart published in 1954, 'Alcohol addiction' has become 'alcohol dependence' and the revised version allows the possibility that the alcoholic can stop drinking and recover before 'rock bottom' has been reached.

13. Edwards, G. and Gross, M. (1976). Alcohol dependence: provisional description of a clinical syndrome. *British Medical Journal*, **1**, 1058–61.

14. See the Report of a Special Committee of the Royal College of Psychiatrists. (1979). *Alcohol and alcoholism*. Tavistock, London. See also Chap. 2. Edwards, G. (1982). *The treatment of drinking problems: a guide for the health professions*, Grant McIntyre, London.

15. Shaw, S. (1979). A critique of the concept of the Alcohol Dependence Syndrome. *British Journal of Addiction*, **74**, 339–48. See also Shaw, S. (1985). The disease concept of dependence. In *The misuse of alcohol: crucial issues in dependence, treatment and prevention* (ed. N. Heather, I. Robertson, and P. Davies). Croom Helm, London.

16. Edwards, G., Gross, M. M., *et al.* (1977). *Alcohol-related disabilities*. WHO Offset Publication No. 32, Geneva.

4. The evidence

1. Davies, D. L. (1962). Normal drinking in recovered alcohol addicts. *Quarterly Journal of Studies on Alcohol*, **23**, 93–104.

2. Selzer, M. L. and Holloway, W. H. (1957). A follow-up of alcoholics committed to a state hospital. *Quarterly Journal of Studies on Alcohol*, **18**, 98–120.

3. Kendell, R. E. (1965). Normal drinking by former alcohol addicts. *Quarterly Journal of Studies on Alcohol*, **26**, 247–57.

4. Heather, N. and Robertson, I. (1983). *Controlled drinking* (revised edn). Methuen, London. This book contains a detailed review of the evidence relevant to the topic of normal drinking in former alcoholics, as well as a discussion of controlled-drinking theory and treatment practice.

5. Armor, D. J., Polich, J. M., and Stambul, H. B. (1978). *Alcoholism and treatment*. Wiley, New York.

6. Polich, J. M., Armor, D. J., and Braiker, H. B. (1980). *The course of alcoholism: four years after treatment*. Rand Corporation, Santa Monica.

7. A description of all the Sobells' research relevant to this issue will be found in Sobell, M. B. and Sobell, L. C. (1978). *Behavioral treatment of alcohol problems*. Plenum Press, New York.

8. Caddy, G. R., Addington, H. J., and Perkins, D. (1978). Individualized behaviour therapy for alcoholics: a third-year independent double-blind follow-up. *Behaviour Research and Therapy*, **16**, 345–62.

9. Pendery, M. L., Maltzman, I. M., and West, L. J. (1982). Controlled drinking by alcoholics? New findings and a re-evaluation of a major affirmative study. *Science*, **217**, 169–75.

10. For a fuller account see the Postscript in Heather, N. and Robertson, I. (1983). *Controlled drinking* (revised edn). Methuen, London.

11. Edwards, G. (1985). A later follow-up of a classic case series: D. L. Davies's 1962 report and its significance for the present. *Journal of Studies on Alcohol*, **46**, 181–90.

12. See Heather, N. (1989). Controlled drinking treatment: where do we stand today? In *Addictive behaviors: prevention and early intervention* (ed. T. Loberg, W. R. Miller, P. E. Nathan, and G. A. Marlatt). Swets and Zeitlinger, Amsterdam.

13. Nordström, G. and Berglund, M. (1987). A prospective study of successful long-term adjustment in alcohol dependence: social drinking versus abstinence. *Journal of Studies on Alcohol*, **48**, 95–103.

14. Peele, S. (1987). Why do controlled drinking outcomes vary by investigator, by country and by era? A cultural analysis of remission in alcoholism. *Drug and Alcohol Dependence*, **20**, 173–201.

15. Helzer, J. E., Robins, L. N., Taylor, J. R., Carey, K., Miller, R. H., Combs-Orme, T., and Farmer, A. (1985). The extent of long-term moderate drinking among alcoholics discharged from medical

and psychiatric treatment facilities. *New England Journal of Medicine*, **312**, 1678–82.

16. Robertson, I., Heather, N., Dzialdowski, A., Crawford, J., and Winton, M. (1986). A comparison of minimal versus intensive treatment interventions for problem drinkers. *British Journal of Clinical Psychology*, **25**, 185–94.

17. See Heather, N. (1989). Controlled drinking treatment: where do we stand today? In *Addictive behaviors: prevention and early intervention* (ed. T. Loberg, W. R. Miller, P. E. Nathan, and G. A. Marlatt). Swets and Zeitlinger, Amsterdam.

18. Cahalan, D. (1970). *Problem drinkers: a national survey*. Jossey Bass, San Francisco.

19. Clark, W. B. and Cahalan, D. (1976). Changes in problem drinking over a four-year span. *Addictive Behaviours*, **1**, 251–60.

20. Chick, J. and Duffy, J. (1979). Application to the alcohol dependence syndrome of a method of determining the sequential development of symptoms. *Psychological Medicine*, **9**, 313–19.

21. Roizen, R., Cahalan, D., and Shanks, P. (1978). Spontaneous remission among untreated problem drinkers. In *Longitudinal research on drug use* (ed. D. Kandel). Wiley, New York.

22. O'Doherty, F. and Davis, J. (1988). Life events, stress and addiction. In *Handbook of life stress, cognition and health* (ed. S. Fisher and J. Reason). John Wiley, London.

23. Mello, N. K. and Mendelson, J. H. (1965). Operant analysis of drinking habits of chronic alcoholics. *Nature*, **206**, 43–6.

24. Mello, N. K., McNamee, H. B., and Mendelson, J. H. (1968). Drinking patterns of chronic alcoholics: gambling and motivation for alcohol. *Psychiatric research report no. 24*. American Psychiatric Association, Washington D.C.

25. Cohen, M., Liebson, I. A., Faillace, L. A., and Speers, W. (1971). Alcoholism: controlled drinking and incentives for abstinence. *Psychological Reports*, **28**, 575–80.

26. Merry, J. (1966). The 'loss of control' myth. *Lancet*, **4**, 1257–8.

27. See Heather, N. and Robertson, I. (1983). *Controlled drinking* (revised edn), Chap. 3. Methuen, London.

28. Marlatt, G. A., Demming, B., and Reid, J. B. (1973). Loss of control drinking in alcoholics: an experimental analogue, *Journal of Abnormal Psychology*, **81**, 233–41.

29. See Stockwell, T. R., Hodgson, R. J., Rankin, H. J., and Taylor,

C. (1982). Alcohol dependence, beliefs and the priming effect. *Behaviour Research and Therapy*, **20**, 513–22.

30. Ledermann, S. (1956). *Alcool, alcoolisme et alcoolisation*. Institut National d'Études Démographiques, Travaux et Documents, Presses Universitaires de France, Paris.

31. de Lint, J. and Schmidt, W. (1968). The distribution of alcohol consumption in Ontario. *Quarterly Journal of Studies on Alcohol*, **29**, 968—73.

32. For an excellent review of the genetic evidence, see Murray, R. M. and Gurling, H. M. D. (1982). Alcoholism: polygenic influence on a multi-factorial disorder. *British Journal of Hospital Medicine*, April. See also Shields, J. (1977). Genetics and alcoholism. In *Alcoholism: new knowledge and new responses* (ed. G. Edwards and M. Grant). Croom Helm, London.

33. See Goodwin, D. (1976). *Is alcoholism hereditary?* Oxford University Press, New York.

34. See, for example, Section III, Part B, Religion and Ethnicity, in Pittman, D. J. and Snyder, C. R. (eds.) (1962). *Society, culture and drinking patterns*. Wiley, New York.

5. Pros and cons of the disease perspective

1. Jellinek, E. M. (1960). *The disease concept of alcoholism*. Hillhouse Press, New Haven.

2. See Room, R. (1984). Sociology and the disease concept of alcoholism. In *Research advances in alcohol and drug problems*, Vol. 7 (ed. R. J. Gibbons *et al.*). Plenum Press, London.

3. Tolor, A. and Tamerin, J. S. (1975). The attitudes towards alcoholism instrument: a measure of attitudes towards alcoholics and the nature and causes of alcoholism. *British Journal of Addiction*, **70**, 223–31.

4. Mulford, H. and Miller, D. (1964). Measuring public acceptance of the alcoholic as a sick person. *Quarterly Journal of Studies on Alcohol*, **25**, 314–23.

5. Crawford, J. and Heather, N. (1987). Public attitudes to the disease concept of alcoholism. *International Journal of the Addictions*, **22**, 1129–38.

6. Strong, P. M. (1979). The alcoholic, the sick role and bourgeois

medicine. unpublished ms., Institute of Medical Sociology, Aberdeen.

7. Gerard, D. L. and Saenger, G. (1966). *Out-patient treatment of alcoholism: a study of outcome and its determinants*. University of Toronto Press, Toronto.

8. Miller, W. R. (1983). Motivational interviewing with problem drinkers. *Behavioural Psychotherapy*, **11**, 147–72.

9. Heather, N., Rollnick, S., and Winton, M. (1983). A comparison of objective and subjective measures of alcohol dependence as predictors of relapse following treatment. *British Journal of Clinical Psychology*, **22**, 11–17.

10. Beauchamp, D. (1980). *Beyond alcoholism*. Temple University Press, Philadelphia.

11. Dight, S. E. (1976). *Scottish drinking habits*. HMSO, London.

12. Sanchez-Craig, M., Annis, H., Bornet, A., and McDonald, K. (1984). Random assignment to abstinence and controlled drinking: evaluation of a cognitive-behavioral program for problem drinkers. *Journal of Consulting and Clinical Psychology*, **52**, 390–403. See also Sanchez-Craig, M. and Lei, H. (1986). Disadvantages of imposing the goal of abstinence on problem drinkers: an empirical study. *British Journal of Addiction*, **81**, 505–12.

13. Gerard, D. L., Saenger, G., and Wile, R. (1962). The abstinent alcoholic. *Archives of General Psychiatry*, **6**, 83–95.

14. For example, Pattison, E. M. (1976). A conceptual approach to alcoholism treatment goals. *Addictive Behaviours*, **1**, 177–92.

15. Davies, P. (1979). Motivation, responsibility and sickness in the psychiatric treatment of alcoholism. *British Journal of Psychiatry*, **134**, 449–58.

16. For a discussion of these issues, see Conrad, P. and Schneider, J. W. (1980). *Deviance and medicalization: from badness to sickness*, Chap. 4. C. V. Mosley, St Louis.

17. Fingarette, H. (1970). The perils of *Powell*: in search of a factual foundation for 'the disease concept of alcoholism'. *Harvard Law Review*, **83**, 793–807.

18. Kendell, R. E. (1979). Alcoholism: a medical or political problem? *British Medical Journal*, **288**, 367–71.

19. Department of Health and Social Security (1981). *Drinking sensibly*. HMSO, London.

6. Setting the scene

1. See Edwards, G., Gross, M., *et al*. (1977). *Alcohol-related-disabilities*, WHO Offset Publication No. 32. WHO, Geneva.

2. Wray, I. and Dickerson, M. (1981). Cessation of high frequency gambling and 'withdrawal' symptoms. *British Journal of Addiction*, **76**, 401–5.

3. Cited in Room, R. (1980). Treatment seeking populations and larger realities. In *Alcoholism treatment in transition* (ed. G. Edwards and M. Grant). Croom Helm, London.

4. See Thorley, A. (1985). The limitations of the alcohol-dependence syndrome in multidisciplinary service development. In *The misuse of alcohol: crucial issues in dependence, treatment and prevention* (ed. N. Heather, I. Robertson, and P. Davies). Croom Helm, London.

5. See Edwards, G., Chandler, J., *et al*. (1972). Drinking in a London suburb: II. Correlates of trouble with drinking among men. *Quarterly Journal of Studies on Alcohol*, Suppl. **6**, 94–119; and Armor, D. J., Polich, J. M., and Stambul, H. B. (1978). *Alcoholism and treatment*. Wiley, New York. Similar limits were arrived at from a medical perspective by Paton, A. and Saunders, B. (1981). ABC of alcohol: definitions. *British Medical Journal*, **293**, 1248–50.

6. Sabey, B. and Staughton, G. (1980). *The drinking road user in Great Britain*, Transport and Road Laboratory, Supplementary Report No. 616. Department of Transport, Crowthorne, Berks.

7. See Marx, M. and Hillix, W. (1963). *Systems and theories in psychology*. McGraw-Hill, New York.

8. Thorndike, E. L. (1905). *The elements of psychology*. A. G. Seiler, New York.

9. Sasmor, R. M. (1966). Operant conditioning of a small-scale muscle response. *Journal of the Experimental Analysis of Behaviour*, **9**, 69–85.

10. Bandura, A. (1969). *Principles of behavior modification*. Holt, Rinehart and Winston, New York.

11. Thoresen, C. E. and Mahoney, M. J. (ed.) (1974). *Behavioral self-control*. Holt, Rinehart and Winston, New York.

12. Schachter, S. and Singer, J. (1962). Cognitive, social and physiological determinants of emotional state. *Psychological Review*, **69**, 379–99.

7. Conditioning

1. Subkov, A. A. and Zilov, G. N. (1937). The role of conditioned reflex in the origin of hypergic reactions. *Bulletin de Biologie et Médecine Expérimentale de l'URSS*, **4**, 294–6.

2. Siegel, S. (1975). Conditioned insulin effects. *Journal of Comparative Physiology and Psychology*, **89**, 189–99.

3. This evidence is reviewed in Hinson, R. E. and Siegel, S. (1980). The contribution of Pavlovian conditioning to ethanol tolerance and dependence. In *Alcohol tolerance and withdrawal* (ed. H. Rigter and J. Crabbe). Elsevier/North Holland Biological Press, New York.

4. Robins, L., Davis, D., and Goodwin, D. (1974). Drug use by US Army enlisted men in Vietnam: a follow-up on their return home. *American Journal of Epidemiology*, **99**, 235–49.

5. Heather, N. and Robertson, I. (1983). *Controlled drinking* (revised edn). Methuen, London.

6. The evidence for this assertion is contained in the references given in Note 5, Chapter 6.

7. Nathan, P. and O'Brien, J. (1971). An experimental analysis of the behavior of alcoholics and nonalcoholics during prolonged experimental drinking: a necessary precursor of behaviour therapy? *Behaviour Therapy*, **2**, 455–76.

8. Cappell, H. and Herman, C. (1972). Alcohol and tension reduction: a review. *Quarterly Journal of Studies on Alcohol*, **33**, 33–47.

9. Stockwell, T., Hodgson, R., and Rankin, H. (1982). Tension reduction and prolonged alcohol consumption. *British Journal of Addiction*, **77**, 65–73.

10. Blakey, R. and Baker, R. (1980). An exposure approach to alcohol abuse. *Behaviour Research and Therapy*, **18**, 319–26.

11. Hodgson, R. J. and Rankin, H. J. (1976). Modification of excessive drinking by cue exposure. *Behaviour Research and Therapy*, **14**, 305–7. See also Rankin, H., Hodgson, R., and Stockwell, T. (1983). Cue exposure and response prevention with alcoholics: a controlled trial. *Behaviour Research and Therapy*, **21**, 435–46. This research is much more complex than could possibly be conveyed here and the interested reader should turn to the original articles above.

12. Orford, J. and Edwards, G. (1977). *Alcoholism*. Maudsley Monograph No. 26. Oxford University Press, London.

13. Smith, M. J. (1975). *When I say no, I feel guilty*. Bantam Books, London.

14. Alexander, J. and Parsons, B. (1973). Short-term behavioural intervention with delinquent families: impact on family process and recidivism. *Journal of Abnormal Psychology*, **80**, 219–25.

15. Hunt, G. and Azrin, N. (1973). A community-reinforcement approach to alcoholism. *Behaviour Research and Therapy*, **11**, 91–104. See also Azrin, N. (1976). Improvements in the community reinforcement approach to alcoholism. *Behaviour Research and Therapy*, **14**, 339–48.

16. Chick, J. (1982). Do alcoholics recover? *British Medical Journal*, **285**, 3–4.

8. Cognition

1. For a general overview see Orford, J. and Harwin J. (1983). *Alcohol and the family*. Croom Helm, London; in particular, the chapter by Claire Wilson. For a study relating childhood characteristics to adult problem drinkers, see McCord, W., McCord, J., and Gudeman, J. (1960). *Origins of alcoholism*. Stanford University Press, Stanford, California.

2. Caudill, B. D. and Marlatt, G. A. (1975). Modelling influences in social drinking: an experimental analogue. *Journal of Consulting and Clinical Psychology*, **43**, 405–15.

3. McAndrew, C. and Edgerton, R. (1969). *Drunken comportment: a social explanation*. Aldine, Chicago.

4. Bandura, A. and Simon, K. (1977). The role of proximal intentions in self-regulation of refractory behaviour. *Cognitive Therapy and Research*, **1**, 177–93.

5. Hodgson, R. (1980). Treatment strategies for the early problem drinker. In *Alcoholism treatment in transition* (ed. G. Edwards and M. Grant). Croom Helm, London.

6. See, for example, Robertson, I. and Heather, N. (1983). *So you want to cut down your drinking?* Scottish Health Education Group, Edinburgh.

7. Mahoney, M. and Arnkoff, D. (1978). Cognitive and self-control

therapies. In *Handbook of psychotherapy and behavior change* (ed. S. Garfield and A. Bergin). Wiley, New York.

8. Hull, J. (1981). A self-awareness model of the causes and effects of alcohol consumption. *Journal of Abnormal Psychology*, **90**, 586–600.

9. Davies, J. and Stacey, B. (1972). *Teenagers and alcohol*. HMSO, London.

10. Aitken, P. (1978). *Ten- to fourteen-year olds and alcohol*. HMSO, Edinburgh.

11. Cahalan, D., Cisin, I., and Crossley, H. (1969). *American drinking practices*. Monograph No. 6, Rutgers Center for Alcohol Studies, New Brunswick.

12. O'Connor, J. *The young drinkers*. Tavistock, London.

13. Briddell, D., Rimm, D., Caddy, G., *et al.* (1978). Effects of alcohol and cognitive set on sexual arousal to deviant stimuli. *Journal of Abnormal Psychology*, **87**, 418–30.

14. Rollnick, S. and Heather, N. (1982). The application of Bandura's self-efficacy theory to abstinence-oriented alcoholism treatment. *Addictive Behaviors*, **7**, 243–50.

15. Seligman, M. (1975). *Helplessness*. Freeman, San Francisco.

16. Marlatt, G. A. (1983). *Relapse prevention*. Guilford Press, New York.

17. D'Zurilla, J. and Goldfried, M. (1971). Problem solving and behaviour modification. *Journal of Abnormal Psychology*, **78**, 107–26.

18. Spivak, G., Platt, J., and Shure, M. (1976). *The problem-solving approach to adjustment*. Jossey Bass, San Francisco.

19. Beck, A. T. (1976). *Cognitive therapy and the emotional disorders*. International Universities Press, New York.

20. Ellis, A. (1962). *Reason and emotion in psychotherapy*. Lyle Stuart, New York.

21. Mischel, W. (1974). Processes in the delay of gratification. In *Advances in experimental social psychology*, Vol. 7 (ed. L. Berkowitz). Academic Press, New York.

22. Chaney, E., O'Leary, M., and Marlatt, G. A. (1978). Skill training with alcoholics. *Journal of Consulting and Clinical Psychology*, **46**, 1092–104.

23. Robertson, I. and Heather, N. (1982). An alcohol education course for young offenders: a preliminary report. *British Journal on Alcohol and Alcoholism*, **17**, 32–8.

24. Mai, N. (1975). Mathematical model for evaluation of therapies. In *Progress in behavior therapy* (ed. J. Brengelmann). Springer, New York.

25. Goldman, M. S., Williams, D. L., and Klisz, D. (1983). Recoverability of psychological functioning following alcohol abuse: prolonged visual-spatial dysfunction in older alcoholics. *Journal of Consulting and Clinical Psychology*, **51**, 370–78.

26. In addition to Marlatt (note 16 above), see the following overviews of behavioural methods: Marlatt, G. and Nathan, P. (ed.) (1978). *Behavioral approaches to alcoholism*. Rutgers Center for Alcohol Studies, New Brunswick; Hay, W. and Nathan, P., (ed.) (1982). *Clinical case studies in the behavioral treatment of alcoholism*. Rutgers Center for Alcohol Studies, New Brunswick; and Miller, P. (1976). *Behavioral treatment of alcoholism*. Pergamon, New York.

9. Some practical implications

1. This is elaborated in Heather, N. and Robertson, I. (1983). *Controlled drinking* (revised edn). Methuen, London.

2. Piispa, M. (1981). Suomen sanomalehdiston alkoholikirjoittelun linjat 1951–1978 (From temperance policy to alcohol education: the Finnish press and alcohol policy 1951–1978). *Alkoholipolitiikka*, **46**, 174–83.

3. Emrick, C. D. (1975). A review of psychologically oriented treatment of alcoholism: II. The relative effectiveness of different treatment approaches and the effectiveness of treatment versus no treatment. *Quarterly Journal of Studies on Alcohol*, **36**, 88–108.

4. Orford, J. and Edwards G. (1977). *Alcoholism*, Maudsley Monograph No. 26. Oxford University Press, London.

5. Hedberg, A. and Campbell, L. (1974). A comparison of four behavioural treatment approaches to alcoholism. *Journal of Behaviour Therapy and Experimental Psychiatry*, **5**, 251–6.

6. See Steinglass, P., Weiner, S., and Mendelson, J. (1971). A systems approach to alcoholism: a model and its clinical application. *Archives of General Psychiatry*, **24**, 401–10.

7. For references to useful overviews of behavioural methods, see note 26, Chapter 8.

8. Cartwright, A. (1981). Are different therapeutic perspectives

important in the treatment of alcoholism? *British Journal of Addiction*, **76**, 347–61.

9. Parker, M., Winstead, D., Willi, F., and Fisher, P. (1979). Patient autonomy in alcohol rehabilitation: II. Program evaluation. *International Journal of the Addictions*, **14**, 1177–84.

10. Sanchez-Craig, M., Annis, H., Bornet, A., and McDonald, K. (1984). Random assignment to abstinence and controlled drinking: evaluation of a cognitive-behavioral program for problem drinkers. *Journal of Consulting and Clinical Psychology*, **52**, 390–403.

11. Robertson, I. and Heather, N. (1982). A survey of controlled drinking treatment in Britain. *British Journal on Alcohol and Alcoholism*, **17**, 102–5.

12. Boffey, P. (1983). Controlled drinking gains as a treatment in Europe. *New York Times*, 22 November 1983.

13. Baekeland, F. and Lundwall, L. (1977). Engaging the alcoholic in treatment and keeping him there. In *The Biology of alcoholism*, Vol. V (ed. B. Kissin and H. Begleiter). Plenum Press, New York.

14. Davies, P. (1979). Motivation, responsibility and sickness in the psychiatric treatment of alcoholism. *British Journal of Psychiatry*, **134**, 449–58.

15. *Mental Health (Amendment) Act* (1982), Chapter 51. HMSO, London.

16. Robertson, I. and Heather, N. (1985). *So you want to cut down your drinking?* Scottish Health Education Group, Edinburgh. A report on the effectiveness of the manual is given in Heather, N., Whitton, B., and Robertson, I. (1986). Evaluation of a self-help manual for media-recruited problem drinkers: six-month follow-up results. *British Journal of Clinical Psychology*, **25**, 19–34.

17. Heather, N., Robertson, I., MacPherson, B., Allsop, S., and Fulton, A. (1987). The effectiveness of a controlled-drinking self-help manual: one year follow-up results. *British Journal of Clinical Psychology*, **26**, 279–87.

18. Chick, J., Lloyd, G., and Crombie, E. (1985). Counselling problem drinkers in medical wards: a controlled study. *British Medical Journal*, **290**, 956–67.

19. See Babor, T. F., Treffardier, M., Weill, J., *et al.* (1983). The early detection and secondary prevention of alcoholism in France. *Journal of Studies on Alcohol*, **44**, 600–16.

20. Kristensson, H., Ohlin, H., Hulten-Nosslin, M. B., *et al.* (1983). Identification and intervention of heavy drinking in middle-aged

men: results and follow-up to 24, 60 months of long term study with randomized control. *Alcoholism: Clinical and Experimental Research*, **7**, 203–9.

21. Scottish Health Education Group, *DRAMS kit for general practitioners*, SHEG, Edinburgh. See also Heather, N., Campion, P., Neville, R., and Maccabe, D. (1987). Evaluation of a controlled drinking minimal intervention for problem drinkers in general practice (the DRAMS Scheme). *Journal of the Royal College of General Practitioners*, **37**, 358–63.

22. Wallace, P., Cutler, S., and Haines, A. (1988). Randomised controlled trial of general practitioner intervention in patients with excessive alcohol consumption. *British Medical Journal*, **297**, 663–8.

23. Miller, W. R. (1983). Motivational interviewing with problem drinkers. *Behavioural Psychotherapy*, **11**, 147–72.

24. Prochaska, J. O. and DiClemente, C. C. (1986). Toward a comprehensive model of change. In *Treating addictive behaviors: processes of change* (ed. W. R. Miller and N. Heather). Plenum, New York.

25. Marlatt, G. A. and Gordon, J. R. (ed.) (1985). *Relapse prevention: maintenance strategies in the treatment of addictive behaviors*. Guilford, New York.

26. Central Policy Review Staff. *Alcohol policies in the United Kingdom* (published in Sweden by the Sociologiska Institutionen, Stockholm).

27. Department of Health and Social Security (1981). *Drinking sensibly*. HMSO, London.

28. Kendell, R., de Roumanie, M., and Ritson, B. (1983). Effect of economic changes on Scottish drinking habits 1978–1982. *British Journal of Addiction*, **75**, 365–80.

29. Scottish Home and Health Department (1973). *Report of the Departmental Committee on Scottish Licensing Laws*. SHHD, Edinburgh.

30. Jeffs, B. and Saunders, W. (1983). Minimizing alcohol-related offences by enforcement of the existing licensing legislation. *British Journal of Addiction*, **78**, 67–77.

31. O'Connor, J. (1978). *The young drinkers*. Tavistock, London.

32. Plant, M. (1979). *Drinking careers*. Tavistock, London.

33. Health and Safety Executive (1981). *The problem drinker at work*. HMSO, London.

34. Dunkin, W. (1981). Policies in the United States. In *Alcohol problems in employment* (ed. B. Hore and M. Plant). Croom Helm, London.
35. Department of the Environment (1976). *Drinking and driving: the report of the Departmental Committee*. HMSO, London.
36. Ritson, B., de Roumanie, M., and Kendrick, S. (1981). *Community response to alcohol problems*. WHO, Geneva.
37. Huesman, L. R., Eron, L. D., *et al.* (1973). 'Television, violence and aggression: the causal effect remains. *American Psychologist*, **28**, 617–20.

Epilogue: the way ahead

1. Crawford, A., Plant, M. A., Kreitman, N., and Latcham, R. W. (1985). Self-reported alcohol consumption and adverse consequences of drinking in three areas of Britain. *British Journal of Addiction*, **80**, 421–8.
2. The address of AAA is: 11 Carteret Street, London SW1H 9DL (Director: Don W. Steele). The other national organization in the field, funded in England and Wales largely by the DHSS, is *Alcohol Concern*, 305 Gray's Inn Road, London WC1X8QF (Director: Dianne Hayter).
3. Dight, S. (1978). *Scottish drinking habits*. HMSO, London.
4. Department of Health and Social Security (1978). *The pattern and range of services for problem drinkers*, Report by the Advisory Committee on Alcoholism. HMSO, London.
5. See Heather, N. and Robertson, I. (1983). *Controlled drinking* (revised edn). Methuen, London.
6. Kalb, M. and Propper, M. S. (1976). The future of alcohology: craft or science? *American Journal of Psychiatry*, **133**, 641–5.

Index